Mastering Autodesk®
VIZ Render
A Resource for Autodesk®
Architectural Desktop Users

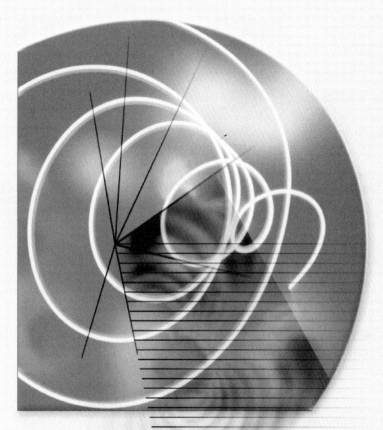

Mastering Autodesk®
VIZ Render
A Resouce for Autodesk®
Architectural Desktop Users

PAUL F. AUBIN
JAMES D. SMELL

Autodesk®

THOMSON

DELMAR LEARNING™

Australia • Canada • Mexico • Singapore • Spain • United Kingdom • United States

Mastering Autodesk® VIZ Render
A Resource for Autodesk®
Archetectural Desktop Users
Paul F. Aubin, James D. Smell

Autodesk Press Staff:

Vice President, Technology and Trades SBU:
Dave Garza

Director of Learning Solutions:
Sandy Clark

Senior Acquisitions Editor:
James DeVoe

Senior Product Manager:
John Fisher

Editorial Assistant:
Tom Best

Channel Manager:
Dennis Williams

Production Director:
Mary Ellen Black

Production Manager:
Andrew Crouth

Senior Production Editor:
Stacy Masucci

Library of Congress Cataloging-in-Publication Data:
ISBN: 1-4180-3963-2

NOTICE TO THE READER

CONTENTS

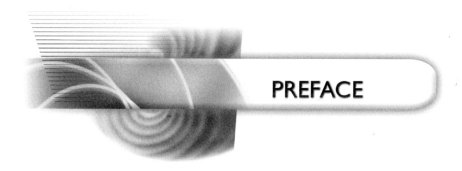

PREFACE

WELCOME

VIZ Render is a standalone rendering and animation package bundled with Autodesk Architectural Desktop. It shares much of its interface in common with its siblings Autodesk VIZ and 3ds max, but has a much more limited feature set than these applications. It is a feature of Architectural Desktop and as such, all modeling (and much material application) is accomplished in Autodesk Architectural Desktop (ADT). VIZ Render (VIZr) is then used to fine-tune the material application, add lights and cameras, and generate high-quality renderings and animations. The goal of this book is to familiarize both novice and experienced 3D artists with the process and capabilities of the ADT/VIZr tool set. The book takes a project-based approach to the exploration of the process of working with ADT and VIZr. In Section I, there are two introductory chapters. Beyond that, in Section II, each chapter will showcase a common project scenario. The process to achieve the stated results will be outlined and a final result created in each of these project scenarios. The overall goal of the book is to give the reader several proven tips and procedures to achieve high quality results.

WHO SHOULD READ THIS BOOK?

The primary audience of this book is any ADT user who wishes to learn about VIZ Render. No VIZr experience is required or assumed. Only beginning level experience in ADT is required. If you are an architect, interior designer, or other building professional that is using Architectural Desktop, and are interested in learning about the benefits, features, and techniques of VIZ Render, then this book is for you.

FEATURES OF THIS EDITION

This book is organized into three sections. Section I includes introductory chapters that cover the basics of VIZ Render usage and interface. The process required to use the software successfully is outlined in this section. In Section II, each chapter follows a project format. A dataset (from the included CD ROM) is loaded and a particular rendering task is outlined as the goal of the chapter. The chapter walks the reader through the tasks required to achieve the stated goals of the chapter in tutorial form. The final section includes Appendices and a Color Plates section. Within the Color section are example renderings from each chapter.

This book is written for users of Architectural Desktop. The text is written primarily for 2006, but the concepts can be applied to VIZ Render 2005 and 2004 as well. If you are using VIZ Render 2005 or 2004, you may want to purchase the first edition of Mastering VIZ Render which covers those releases specifically.

STYLE CONVENTIONS

Style Conventions used in this text are as follows:

Text	Autodesk Architectural Desktop.
2. Step-by-Step Tutorials	Perform these steps.
Commands and Menu Picks	**Rendering > Render**
Keyboard input	type **COMMAND** and then press ENTER.
User Interaction with interface elements	right-click, and choose **Move** from the Edit/ Transform Quad menu.
File and Directory Names	C:\MasterVIZr\Chapter02\Sample File.drf

 Features new to ADT 2006 are indicated by this icon.

HOW TO USE THIS BOOK

While you are not required to read all the chapters in order, it is recommended that you read Chapters 1 and 2 before attempting any of the projects. Chapters 3 and 4 cover modeling in ADT. If you wish to start at the beginning of the modeling process, read Chapter 3 first, then 4. If you are already familiar with the basics of building models in ADT, skip ahead to Chapter 4. If you would rather get right into rendering, skip right to Chapter 5. Chapters 5 through 9 can be read in order, or you can skip around to the project you find most interesting if you wish.

HOW TO USE THE CD

Files used in the tutorials throughout this book are located on the included CD ROM. The required files for each chapter have been provided so that you can start at the beginning of any chapter without worry that a required file(s) will be missing. Also included are the final renderings from each chapter. Therefore, you will be able to load the file for a given chapter and begin working without being required to finish the previous chapter first. When you install the files from the CD, the files for all chapters will be installed automatically. The files will install into a folder on your C: drive named *MasterVIZr*. Inside this folder will be a folder for each chapter—for example, files for Chapter 5 will be found in a folder named *Chapter05*. In addition to the nine chapter folders, there is also a folder for each *Appendix, Maps, Renderings, Render Presets,* and several others. All items are referenced to their installed location within the text of the exercises.

Installing CD Files

Locate the Mastering VIZ Render CD ROM in the back cover of your book. (Read the license agreement and be sure that you agree to its terms before you break the seal—once you have broken the seal to the CD, you cannot return the book.) To install the dataset files, do the following:

1. Place the CD ROM in your CD drive.

2. An installer window should appear onscreen after a moment or two.

3. Click the Install MasterVIZr Dataset button.

Installation will commence automatically and all files will be installed to a folder named *C:\MasterVIZr* on your hard drive.

4. Click **Install Appendix C Files** to install the Appendix C Files.

Installation will commence automatically and all files will be installed to a folder named *C:\MasterVIZr\Appendix C* on your hard drive.

 CAUTION *If you choose to move the files from this location the Tool Catalogs included within may not function properly. Moving any of the other files can also cause issues with ADT project files.*

If you do not intend to perform the tutorials in certain chapters, it is OK to delete the files for those chapters. Simply delete the entire folder for the chapter(s) that you wish to skip. Complete installation requires approximately 480 MB of disk space on your C: Drive.

 NOTE *If you have a previous version of mastering VIZ Render, please rename or delete the existing Master VIZr folder on your C: Drive befor eintsalling the new one.*

MUSIC SELECTIONS

When you perform renderings in VIZ Render (or any 3D Rendering software) they can take quite a long time to complete. To help you pass the time, we have suggested selected music to play while you wait for your renderings to complete. Naturally we have tried to pick songs that last about the duration of the rendering, but this is not an exact science and the times will vary based upon a number of factors. Within the text of each chapter, after each topic that instructs you to render, you will see something like the following:

 Render Stats 01.03

Render Time: 4 minutes 21 seconds
Radiosity Processing Time: 1 minute 6 seconds
Music Selection: "In These Shoes" Bette Midler – Bette (2000)
Color Plate: C–5
JPG File: Rendering 05.01.jpg

If you have this particular music selection in your personal library, feel free to pop it in while you wait for your rendering. If you do not like the music we selected, please feel free to choose a selection that better suits your own personal tastes. There is a lot of waiting involved in rendering. The music selections are simply a fun way to pass the time. If you prefer, feel free to whip up a smoothie, take the dogs for a walk, or break out the *New York Times* crossword puzzle—whatever you like to do to pass the time. If you like the idea of the music selections that we have provided, the complete list is available as a playlist on the Apple Music Store. Visit *www.paulaubin.com* for a link to this playlist. If you would like to make requests for future playlists, you can do this at *paulaubin.com* as well. You can download this playlist and, if you wish, purchase any of these songs from the Apple iTunes Store and/or Amazon.com. Please note, we do *not* include the songs on the CD. You will need to purchase them if you do not have them already in your personal music library.

PATHS

To be sure that all of the resources provided on the Mastering VIZ Render CD ROM function properly, you will need to add a path statement in VIZr and in ADT. To do this:

1. Launch ADT and then from the Format menu, choose **Options**.

2. On the Files tab, expand Texture Maps and then click the Add button.

3. Browse to the *C:\MasterVIZr\Maps* folder and then click Open.

4. Close the Options dialog and then Exit ADT.

5. Launch VIZr and then from the Customize menu, choose Configure Paths.

 NOTE If a dialog appears asking you to select a Video Driver, choose Software and then click OK. If you are certain about one of the other choices instead, you can choose it instead of Software. Check with your IT or CAD Support Personnel first.

6. On the External Files tab, click the Add button.

7. Browse to the *C:\MasterVIZr\Maps* folder and then click Use Path.

8. Close the Configure Paths dialog and then Exit VIZ Render.

THIRD-PARTY PLUGINS

Two third-party plugins are included with Architectural Desktop for use in VIZ Render, but must be installed before they can be used. Installation files can be found on the Mastering VIZ Render CD ROM in the *C:\MasterVIZr\Plugins* folder. To be sure that you are installing the latest version of each plugin, visit the manufacturer's website.

The two plugins that we will be using in some of the tutorials in this book include the following:

Bionatics® EASYnat—Please refer to the Bionatics website for more information: *www.bionatics.com.*

ArchVision® RPC™—After installing, the RPC content path will need to be added to the list of External Files paths in the Configure Paths dialog. This dialog can be found on the Customize pulldown. Please refer to the ArchVision website for more information: *www.archvision.com.*

If you choose not to download and install these plugins, you will still be able to follow the tutorials in this book. However, you will not be able to complete the specific topics that reference these tools.

ONLINE

Be sure to utilize the VIZ Render Online Help system and take the time to install the online tutorials that come with the software. Both of these resources will provide you with additional help and support as you gain comfort and familiarity with VIZ Render.

WE WANT TO HEAR FROM YOU

If you have any questions and comments about this text, please contact:

The CADD Team
c/o Autodesk Press
Executive Woods
5 Maxwell Drive
Clifton Park, NY 12065-8007
Website: *www.autodeskpress.com*

ABOUT THE AUTHORS

Paul F. Aubin has a background in the architectural profession spanning over 15 years. In addition to writing, Paul travels the country speaking and providing implementation, training, and support services. He currently serves as the Moderator for *Cadalyst* magazine's online CAD questions forum, and has been published in their print magazine.

He is a regular speaker at Autodesk University® and was the recipient of the Autodesk® Central Region 2001 Architectural Award of Excellence. Paul has been a guest speaker at Chicago area events for The American Institute of Architects Northeast Illinois, Chicago Area Users' Groups, and The Association of Licensed Architects. Prior to becoming an independent consultant, Paul was an Associate with a leading Autodesk Systems Center in Chicago where he successfully trained over 1,100 professionals in Architectural Desktop, AutoCAD, and 3D Studio VIZ. While in architectural practice, Paul served as CAD Manager with an interior design and architecture firm in downtown Chicago, and amassed many years of hands-on architect-level project experience. The combination of his experiences in architectural practice, as a CAD manager, and an instructor give his writing, classroom instruction, and consultation a fresh and credible focus. Paul is an associate member of the American Institute of Architects. He received his Bachelor of Science in Architecture and his Bachelor of Architecture from The Catholic University of America. Paul lives in the southwest suburbs of Chicago with his wife, Martha, their sons, Marcus and Justin, and daughter, Sarah Gemma.

You can visit Paul's web site at *www.paulaubin.com*.

James D. Smell pursued study abroad with the Atelier Veneziano, the University of Kentucky College of Architecture's Venice Program. He received a Bachelor of Architecture from the University of Kentucky and is a registered Architect (AR0017182) and an active member of the American Institute of Architects. He has held instructional positions at Jacksonville University as an adjunct professor and is a guest instructor at Florida Community College at Jacksonville. While in Jacksonville, Jim successfully trained over 500 professionals and students in Autodesk VIZ, Architectural Desktop, and AutoCAD. Prior to training as an Autodesk Certified Instructor, Jim worked in architectural firms in Kentucky and Florida, where he performed the duties of CAD Manager and Project Manager. Jim lives in Manchester, New Hampshire, with his wife, Kelly, his daughter, Taryn, and son, Liam. He is currently employed by Autodesk in the Building Solutions Division.

You can send email comments and questions on the content of this book directly to the authors at: *MasterVIZr@paulaubin.com*.

DEDICATION

Paul's Dedication: This book is dedicated to my son Justin. Thanks kiddo for being Daddy's little helper

Jim's Dedication: This book is dedicated to my wife Kelly Collins who never ceases to amaze me with her patience, love and understanding. Thank you.

ACKNOWLEDGEMENTS

The authors would like to thank several people for their assistance and support throughout the writing of this book.

Thanks to Jim Devoe, John Fisher, and all of the Delmar team. It continues to be a pleasure to work with so dedicated a group of professionals.

Thanks to Eric Stenstrom for technical editing, Rachel Pearce Anderson for copy editing, and John Shanley of Phoenix Creative Graphics for composition.

The following individuals provided invaluable technical assistance in the writing of this book. The authors would like to extend special thanks to:

Jon Anderson – Archvision, Rob Finch – Autodesk BSD, Jon Rossen – Discreet, Mark Webb – Autodesk BSD (Thanks, Mark, for your assistance on our i-drop pages for the CD), Steven Papke – Vizdepot, Beau Turner – Vizdepot, and especially Alexander Bicalho – Discreet (Thanks, Alex, for tirelessly answering our endless stream of questions, and a special thanks forfor updating the installation of Appendix C).

We are very grateful to the following vendors and content providers for generously granting permission to use their content in these pages and provide samples of their content on the CD ROM:

Randall Stevens – Archvision, François Lebled – Bionatics, Christine Morse – Herman Miller, Scott Onstott – i-drop content, and Martin Krautter – ERCO.

There are far too many folks in Autodesk Building Solutions Division (BSD) to mention. Thanks to all of them, but in particular, Bryan Otey, Julian Gonzalez, Chris Yanchar, Paul McArdle, Bill Glennie, Dennis McNeal, William (Fitz) Fitzpatrick, Jim Awe, Kelcy Lemon, Matt Dillon and Scott Reinemann, and all of the folks at Autodesk Tech Support.

Paul would like to extend a special thanks to Don McKenna for all his invaluable assistance. He would also like to extend thanks for his family; parents, Maryann and Del, brothers, Marc and Tom; and his wonderful children, Marcus, Justin, and Sarah Gemma. You three are a source of constant inspiration. Finally, he is most grateful for the constant love and support of his wife, Martha.

Jim would like to extend his personal thanks for the blessings of being surrounded by so many friends, Toby, Rick, and Bridget; and family, his late mother, Karen; his father and stepmother, Tom and Tina; brother, Jeff, and his wife, Cyndy; his entire family of in-laws; and his wonderful children, Taryn and Liam.

Understanding VIZ Render

This section introduces the methodology and interface of Autodesk VIZ Render. In Chapter 1, we'll begin with a quick tutorial-based overview of the software and process. We'll continue in Chapter 2 with an overview of the user interface and many of the most common tools and procedures.

Section I is organized as follows:

Chapter 1   Getting Started

Chapter 2   User Interface and General Overview

Getting Started

INTRODUCTION

In this chapter, we will learn what Autodesk VIZ Render (VIZr) is all about. This chapter is designed to get you acquainted with the VIZr work environment and to explain the basic requirements for its use. We will take a quick tour of VIZr below in tutorial form to see some of the major features of the software.

In our opinion, architecture is about habitable space and how that space affects its inhabitants. Light plays a large role in how people experience space: how we are included or excluded, uplifted and channeled. Viewpoint plays another major role: our approach, the way the space flows, our movement through it. In this chapter we will work with a simple model and investigate the major tools provided in VIZ Render: Daylight, Radiosity, and Cameras.

OBJECTIVES

- Explore the basics of VIZ Render
- Gain comfort with the user interface
- Explore the environment
- Produce your first rendering

VIZ RENDER AND ADT

VIZ Render (VIZr) is an advanced design visualization tool that is integrated into Autodesk Architectural Desktop. It is derived from its parent application Autodesk VIZ, but is optimized and streamlined to perform rendering and simple animation functions only. You cannot create any objects in VIZ Render other than Cameras and Lights. All models are constructed first in Autodesk Architectural Desktop (ADT). These models are then transferred to VIZr using an integrated linking mechanism to create spectacular renderings of these creations.

ADT *is* required to generate renderings from VIZr. VIZ Render is a feature of Autodesk Architectural Desktop and cannot be purchased or used as a separate application. To create a rendering, you must first build a three-dimensional model in ADT. You then link your model to VIZr, where you can add lighting, cameras, edit, or add materials. Finally you generate a rendered image. The results can be simple and diagrammatic, or stunningly photorealistic. (We will look at some sample files shortly.) The results you get out of the software depend highly on what you put into it. Although ADT is required, your level of expertise in ADT can vary widely and still yield good results with VIZr. In other words, it is not necessary for you to master ADT before embarking on learning VIZ Render. A casual familiarity of the basics is sufficient. 3D Rendering is realistically a trinity of modeling, lighting/materials, and rendering. It is naturally valuable to know as much as you can about each of these three separate but equally important parts of the 3D rendering process, however, mastery of each is not a prerequisite to working with the others.

COLOR PLATES

A color plates section is included in this book. Within it you will find 16 pages of color images. The final renderings created in each chapter of this book (and some additional color images) have been included here. Before we begin our exploration of VIZ Render below, please page through that section now to get an idea of the kinds of images we will be creating. If you read the entire book and perform all of its exercises, you will generate your own versions of each of these renderings. Please note that the images that you will see in the color plates section come *directly* from VIZ Render without any post processing. In addition to the color plates section, JPG files have been provided of each rendering produced throughout the book in the *Renderings* folder within the *C:\MasterVIZr* folder. In the Color Plates section in particular, you are encouraged to study the Townhouse Façade rendering (Color Plate C–5 created in Chapter 5), the nighttime Conference Room rendering (Color Plate C–10, created in Chapter 8), and the Orthographic Compilation Presentation (Color Plate C–14, created in Chapter 9).

THE RETREAT DATASET

In each of the forthcoming chapters, we will begin with a brief description of the dataset that we will use in its tutorials. This chapter does not cover modeling. Therefore, the features of the model are described briefly here. If you wish to learn more about modeling techniques right away, you can look ahead to Chapters 3 and 4.

The Retreat project is a simple model. It is a "Box on the Landscape." The model was built using simple ADT objects. A brief description of each of these components follows.

NOTE If you wish to open and explore these components within the model, please feel free to do so. However, be sure to close the file without saving before continuing on to the exercise.

The dataset contains a small retreat building sitting on a relatively flat piece of landscape. The landscape was modeled with a Mass Element Drape object. In ADT you can create terrain models using the Drape tool on the Massing Tool Palette. A Drape is a Mass Element constructed from a collection of polylines that represent the contours; much like a standard contour map, only the polylines must be placed at the proper Z-elevation before generating the Drape. The footprint of the building and the reflecting pool were carved away from the Drape using standard Mass Element primitives and a Boolean Subtract. Additional Mass Elements are used for some of the architectural details such as suspension cabling and door hardware.

The retreat building itself is comprised of Walls, Doors, Windows, and a Space for the floor and ceiling. Slabs are used for the overhangs. Where a simple surface is required, such as the surface of the water in the reflecting pool, AEC Polygons have been used. Other AutoCAD style surfaces can also be used, but since AEC Polygons are ADT objects, they make use of the profile editing features of ADT and, more importantly, they are Style-based. This is important in VIZ Render due to the fact that materials, by default, are applied to all instances of a particular style.

Facet Deviation (FACETDEV) is an ADT setting that controls the smoothness of curved ADT objects. The lower the value, the smoother a curved surface will appear. In this dataset, the FACETDEV setting has been adjusted to 1/16′ from its default value of 1/8′. This is a global setting and helps make all of the ADT objects in this model appear smoother.

Start Autodesk Architectural Desktop and Load the Retreat Dataset

Let's get started using VIZ Render right away. For the next few minutes, we will take the "whirlwind" tour of the VIZr tool set. All of the tools covered in the following steps use default VIZr settings. The chapters that follow cover each of these items and settings in greater detail.

1. If you have not already done so, install the dataset files located on the Mastering VIZ Render CD ROM.

   Refer to "Files Included on the CD ROM" in the Preface for information on installing the sample files included on the CD.

We begin our exercise in Architectural Desktop.

2. Launch Autodesk Architectural Desktop from the icon on your desktop.

You can also launch ADT by choosing the appropriate item (Autodesk Architectural Desktop) from your Windows Start menu in the Autodesk group. In Windows XP, look under "All Programs"; in Windows 2000, look under "Programs."

 NOTE Be sure to load Architectural Desktop, and not VIZ Render, for this step.

3. From the File menu, choose **Open**.

4. Browse to the *C:\MasterVIZr\Chapter01* folder, and then double-click the file named *Retreat.dwg* to open it.

MOVING BETWEEN ADT AND VIZR

Examine the dataset file on screen and notice the existing cameras named "Front Perspective" and "Interior Perspective." These are AEC Camera objects. An AEC Camera is an ADT object that assists you in creating a perspective view from your file. When an AEC Camera is created, an associated AutoCAD Named Model Space View is also created. To "look through" the Camera, you can simply right-click on it, or you can restore one of these associated Views.

Cameras and Views in ADT

Cameras and Views are used to save a particular vantage point in the model. These Views and Cameras can be restored at anytime, optionally including other associated settings such as User Coordinate System (UCS) and Layers. When ADT files are linked over to VIZ Render, the Views automatically translate to VIZr Cameras.

5. Select the Front Perspective Camera, right-click, and choose **Create View**.

6. On the View menu, choose **Shade>Hidden** (see Figure 1–1).

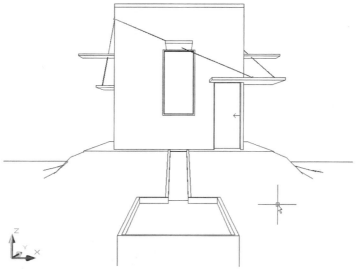

Figure 1–1 *Restore the Front Perspective Camera Model Space View*

As you can see, when you have an AEC Camera object in the drawing, this is the easiest method through which to view it. (If you wish to add your own AEC Camera object, use the tool on the Massing Tool Palette.) However, to see the two Views associated with the Cameras in this file, choose Named Views from the View menu. In the View dialog, you will note the named views "Front Perspective" and "Interior Perspective" matching the names of the AEC Cameras.

7. From the View menu, choose **Named Views** (or type **V** at the command line and then press ENTER).

8. Double-click **Interior Perspective** and then click **OK** to exit the dialog.

9. Open the View dialog again.

Note the additional named Model Space View called: "SW Corner Perspective." This additional named view demonstrates that you do not need an AEC Camera to save a particular view of your model. The AutoCAD Named Views command is sufficient. The AEC Camera simply creates the View for you.

10. Double-click **SW Corner Perspective** and then click **OK** to exit the dialog.

11. Open the View dialog again.

12. Click on the Orthographic & Isometric Views tab. Double-click on **Top** and then click **OK** to exit the dialog.

TIP Preset orthographic Views like Top are also available on the Views toolbar. If the Views toolbar is not loaded, you can right-click any toolbar icon and choose Views from the menu.

Cameras in VIZ Render

Now let's see what happens with these Views and Cameras as we link the file over to VIZ Render. Along the bottom edge of the drawing window is a horizontal band with several controls and readouts. On the left-hand side of the Drawing Status Bar is a small triangle icon. This is the Open Drawing Menu, which is used for many ADT export functions, including linking to VIZ Render (see Figure 1–2).

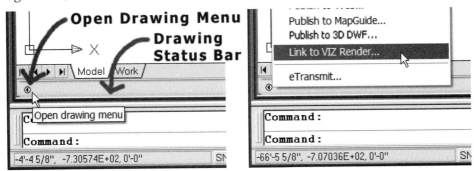

Figure 1–2 *Use the Open Drawing Menu on the Drawing Status Bar to Link to VIZ Render*

13. From the Open Drawing Menu, choose **Link to VIZ Render**.

When you choose to "Link" to VIZ Render, the current ADT drawing will be exported to VIZ Render and linked into a scene there. A Link behaves like an AutoCAD XREF and remains connected, rather than being a one-time export. At any time in VIZ Render, you can choose to reload the Link and retrieve any changes that have occurred in the ADT model.

Let's see now how those Named Model Space Views from ADT have been translated into the VIZr scene.

14. Type **C** on your keyboard.

 NOTE *Like ADT, there are keyboard shortcuts to help maneuver through the VIZr user interface. Chapter 2 will cover some of these in more detail. However, unlike ADT, keyboard shortcuts in VIZr are not followed by the ENTER key.*

Notice that all three of the Named AutoCAD Model Space Views have been imported into the VIZr scene as Cameras; even the one that did not have an associated AEC Camera object in ADT.

15. In the Select Camera dialog, choose: **Camera:Front Perspective**, and then click OK.

Since we will be working with 3D geometry, it is helpful to be able to quickly view the model from multiple angles such as Top, Left, Perspective, etc. For the next several steps, we will be taking advantage of the ability to display a multiple viewport configuration in VIZr. To enable this and to perform other viewport navigation tasks, we will rely heavily on the Viewport Navigation toolbar in this exercise.

16. Click the **Minimize Viewport** icon to minimize the Camera view (see Figure 1–3).

Figure 1–3 *The VIZ Render Viewport Navigation Tools Toolbar contains the primary viewport navigation tools*

This will change the screen to a four-equal viewport configuration. (The viewport layout is highly customizable; more on this in Chapter 2.)

17. On the Viewport Navigation Tools toolbar, click on the **Zoom Extents All** icon (see Figure 1–4).

Figure 1–4 *Use Zoom Extents All to show the maximum view in each viewport*

This will maximize the view in each of the Top, Front, and Left viewports. With this configuration, you can now very clearly see that our building is perched atop a small hill.

Create a New Camera

Now let's move to the inside of our retreat pavilion. For this sequence, we will begin with the existing interior perspective camera angle and then adjust it a bit to create a new Camera.

18. Right-click in the lower-left viewport (currently set to Left) to make it active.

19. On the keyboard, press the **C** key and then choose **Camera:Interior Perspective** in the Select Camera dialog.

The vantage point through this existing camera is a bit high. We can adjust what we see interactively using the Viewport Navigations Tools. Take a look at the Viewport Navigations Tools toolbar. Make a mental note of the various icons (see Figure 1–4). Each of these icons controls some aspect of the currently selected camera view. Camera controls can be tricky to get used to. Sometimes it is easier to first navigate a perspective view and then create a camera from it. We will try that approach here. Switching to a perspective view and later creating a new camera allows you to return to the original in case you make a mistake. This is a good habit to develop.

20. Verify that the Camera:Interior Perspective viewport is active. (If it is not, then right-click anywhere inside to activate it.) Press the **P** key on your keyboard.

This will create a perspective view with the same vantage point as the camera.

21. Right-click on the Perspective viewport title label Perspective, and choose **Smooth + Highlights** (see Figure 1–5).

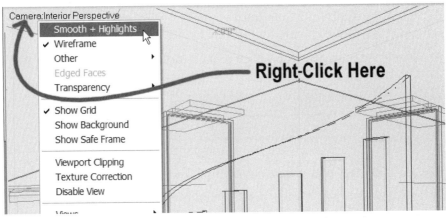

Figure 1–5 *Turn on Smooth + Highlights shading in the Perspective viewport*

Remember that mental note you took above; have another look at the Viewport Navigation Tools Toolbar. Notice that most of the icons have changed. (You can compare Figures 1–4 and 1–6 if short-term memory has failed you.) You get different controls depending on whether you are in a camera, perspective, or orthographic viewport. More information on this can be found in Chapter 2.

Figure 1–6 *The controls on the Viewport Navigation Tools Toolbar adjust to the type of viewport*

22. Press Click the Arc Rotate icon on the Viewport Navigation Tools toolbar.

Depending on where you move your mouse within and around the arc ball you will see one of four cursors. The square handles at the four quadrants of the circle constrain movement in either the vertical or horizontal axes. Using the vertically constrained handles, let's adjust the view to look down on the font in the center of the scene.

23. Move your mouse over the handle at the top of the arc ball, click, and drag down to adjust the view to be looking down on the font in the center of the scene.

    Make your scene match as closely as you can to Figure 1–7 and then right-click to cancel your current Viewport Navigation Tool.

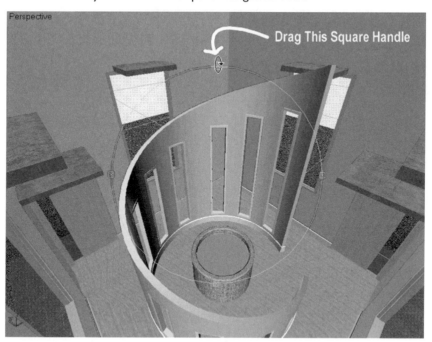

Figure 1–7 *The square handles at the edges of the arc rotate circle constrain movement in one direction*

TIP If you don't like the way a movement is progressing, press the right mouse button before releasing the left button to cancel the operation and return to the view active before the current operation was started.

Now that we have the Perspective View the way we like it, we can convert this viewpoint into a Camera.

24. From the Create menu, choose **Camera > Create Camera From View**.

    The Viewport Label will change to Camera01.

25. From the File menu, choose **Save**.

Reload Geometry

Let's assume that while we were working in VIZ Render preparing our camera views for rendering that one of our colleagues made some changes to the design. They have decided to narrow the Windows. In order to make this change, we must return to ADT and edit the model. Then, to have these changes reflected in our VIZr model we must reload the modified ADT model geometry.

    ADT should still be running with the *Retreat.dwg* file loaded. If you have closed it, re-launch it now and re-open the *Retreat.dwg* file.

26. Select all four exterior Windows, right-click, and choose **Properties**.

TIP To quickly select all four windows, you could select one, right click and then choose "Select Similar" from the context menu

27. On the Properties palette, change the Width to: **2'-6"**.

28. From the File menu, choose **Save**.

29. From the Open Drawing menu (see Figure 1–2 above), choose **Link to VIZ Render**.

    VIZ Render, which should still be running, will become active, and the File Link Settings dialog will appear on screen.

30. In the File Link Settings dialog, simply click OK.

Notice the Windows are now much narrower reflecting the change that we made in the ADT model (see Figure 1–8).

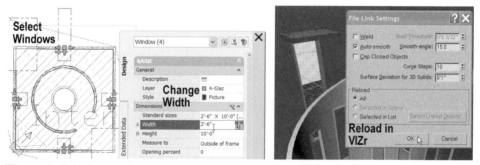

Figure 1–8 *Make an edit to the ADT model and then reload it in VIZ Render*

DAYLIGHT

Our next task will be to create a Daylight System which simulates the light from the Sun. This will be easiest to accomplish in the Top viewport, so let's make it active now. To make a viewport active in VIZr without selecting or deselecting any objects, use the right mouse button.

1. Right-click anywhere inside the **Top** viewport to make it active. (Be sure to use the right mouse button.)

   You should now see a bold outline surrounding the Top viewport.

2. From the Create menu, choose **Daylight System**.

A Daylight Object Creation dialog will appear with a recommendation for Logarithmic Exposure Control settings. The setting offered here works best for exterior daylight renderings. Since we are preparing an interior rendering, we will *not* accept this recommendation.

3. In the Daylight Object Creation dialog, click the **No** button (see Figure 1–9).

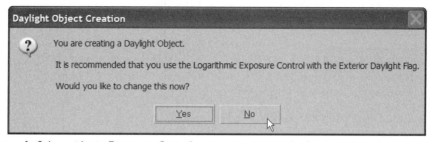

Figure 1–9 *Logarithmic Exposure Control settings prompt in the Daylight Object Creation dialog*

In the next few steps we will be placing the Daylight System. A Daylight System in VIZr consists of a Compass object and a Daylight object. These are representative objects, whose size and location is not critical. Therefore, as you place the

Daylight System, you can size and locate them however you wish. The entire system will be created with a series of clicks and drags. Essentially the steps are: Click – Drag – Release – Drag – Click.

4. **Click** a point in the scene anywhere near the center of the building.

5. **Drag** a little bit to size the Compass.

This drag can be in any direction to increase the size of the Compass. Drag just enough so that the size of the compass is legible. The exact size is not important.

6. **Release** the mouse button when you are happy with the size of the Compass.

7. **Drag** to set the distance above the ground plane of the Daylight object.

This is the trickiest step in this process. The Daylight object will appear to move in a quirky fashion as you drag. Drag the mouse in a vertical motion away from the center of the Daylight System. Do not click until you are happy with its location.

8. **Click** when you are happy with the position of the Daylight object.

 TIP Try picking somewhere around the South edge of the terrain object.

9. From the File menu, choose **Save**. (You can also press CTRL + S).

DRAFT RENDERING

By now you are probably ready for your first rendering! We have provided some "presets" (with the files installed from the Mastering VIZ Render CD ROM) to get you started and then adjust them manually to make improvements in later steps.

Using Render Presets

1. Make sure that Camera01 is the active viewport (if it isn't, right-click in it now), and then choose **Render** from the Rendering menu.

   Locate the Preset dropdown at the bottom of the dialog.

2. From the Preset dropdown list, choose **Load Preset**.

   Browse to C:\MasterVIZr\Render Presets and choose **Draft without Radiosity** (see Figure 1–10).

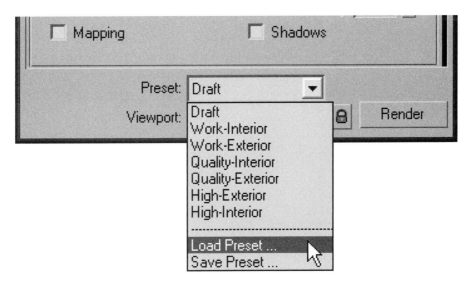

Figure 1–10 *Choose Draft without Radiosity from the default list of Presets*

3. In the Select Preset Categories dialog, be sure that all items are selected and then click the Load button.

4. Click the Render button to the right of the Preset dropdown (see Render Stats 01.01).

Render Stats 01.01

Render Time: 1 minute 26 seconds
Radiosity Processing Time: NA
Music Selection: "Clouds Up" Air – The Virgin Suicides [Original Soundtrack] (2000)
Color Plate: NA
JPG File: Rendering 01.01.jpg

Improving the Render Results

The first rendering yields pretty unimpressive results. It is way too bright, and it rendered way too slowly for a draft rendering. Let's see what we can do to fix this.

5. From the Tools menu, choose **Selection Floater**.

6. On the right side of the Selection Floater, beneath List Types, click the **None** button, and then place a checkmark in the **Groups/Assemblies** checkbox (see Figure 1–11).

This shortens the list of selectable objects to only those that are Groups and Assemblies; in this case, the Daylight System is the only object that qualifies.

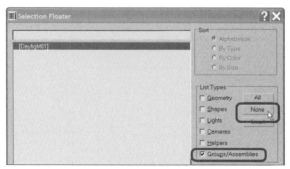

Figure 1–11 *Filter the list of objects to only those that are Groups and Assemblies*

7. Double-click on **[Daylight01]** and then click **Close**.

8. On the Modify panel, in the IES Sky Parameters rollout, select **Partly Cloudy** (see Figure 1–12).

Figure 1–12 *Change the Sky Conditions to Party Cloudy*

9. In the same rollout, change the Rays per Sample to **5**.

This setting relates to how many rays are bounced off a surface. The more there are, the more accurate the results are. Essentially, the Rays per Sample setting controls the accuracy and quality of the daylight solution. However, better results come at the cost of longer rendering time. In one of our tests, a setting of 5 rendered over 3 times faster than a setting of 20. Since we are looking for reasonable results in little time, we'll go for 5 rays here.

10. From the Rendering menu, choose **Environment**.

11. In the Logarithmic Exposure Control Parameters rollout, change the Brightness to **15**.

This setting is like changing the amount of light that enters your eye. When conditions are really bright, your pupil constricts (lower settings). When there isn't enough light, it opens (higher settings). Right now there is too much light, so we've turned it down a bit.

12. At the bottom of the dialog, click the **Render** button (see Render Stats 01.02).

Render Stats 01.02

Render Time: 31 seconds
Radiosity Processing Time: NA
Music Selection: Final Jeopardy Clock Music
Color Plate: NA
JPG File: Rendering 01.02.jpg

This one took much less time and the result is better. However, the light is too even and not very believable. We need to make some more changes.

Higher Quality Presets

Next, let's see the effects of another Render Preset. The nice thing about a Render Preset is that all of the settings are saved with it. So you can easily swap out different collections of settings to suit different goals while you're rendering. In later chapters, when you learn to fully customize Render settings, you can save your own Render Presets.

13. If you closed your Render dialog, open it again. From *C:\MasterVIZr\Render Presets*, choose **Draft with Radiosity**.

14. In the Select Preset Categories dialog, be sure that all items are selected and then click the **Load** button.

Radiosity is a method of rendering based on bounced light, giving you a more believable and realistic rendered image. This Preset turns on Radiosity processing for this rendering. Another thing to note about the Preset is that it changed our Brightness setting back to the default.

15. In the Logarithmic Exposure Control Parameters rollout, change the Brightness to **15** again.

16. Click the **Render** button again (see Render Stats 01.03).

Render Stats 01.03

Render Time: 21 seconds
Radiosity Processing Time: 6 seconds
Music Selection: "Stop" Pink Floyd – The Wall Disk 2 (1979)
Color Plate: NA
JPG File: Rendering 01.03.jpg

The lighting is now very different, but the overall rendering is still not great. You can start to pick up some variances in the light on the room's surfaces.

Understanding the Influence of Radiosity

If you were paying attention to the progress bar during the last rendering, you may have noticed that a "Processing Radiosity" message briefly appeared.

17. In the Render Scene dialog, click on the Radiosity tab and expand the Statistics rollout (see Figure 1–13).

 TIP You may need to scroll to see the Statistics. Do this with the small hand cursor, or the wheel on your mouse.

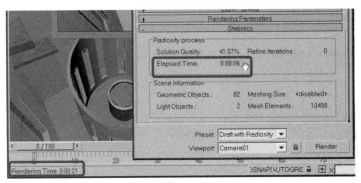

Figure 1–13 *View Render time and Radiosity Statistics*

On our test machine, this rendering took approximately 21 seconds. However, 6 seconds of that were spent processing the Radiosity solution. Your times will naturally vary depending on the specifics of your computer hardware; however, the results should be similar. A significant portion of the total rendering time is consumed by the Radiosity processing. Look at the bottom left corner of your screen on the status bar and you can see the amount of time it took to render your last frame.

18. Click the **Render** button again.

The quality of the rendering is unchanged since we made no changes to the settings. However, if you look again at the Status bar and compare the new rendering time to the previous one, you will note that it is shorter by roughly the amount of time it took to previously process the Radiosity. If you take all your times and do some math, you'll see that once the solution has been calculated, rendering with Radiosity can be significantly shorter than rendering without it. For now let's make a few more changes so we can get to something of acceptable quality.

19. Close the rendering frame and the Render Scene dialog.

20. **Save** the file.

REFINING THE SCENE

We now need to work on making some changes to some of the materials that were assigned to the ADT model, starting with the neon green pool of water that looks like it came from a scene in Smallville™.

Naming Objects

If you are not already in the four viewport setup, please click the Minimize/Maximize icon or press ALT + W.

1. Right-click in the Front viewport to make it active, and if it is not already in Wireframe, right-click on the viewport title and set it to Wireframe.

2. In the Front viewport, use the Zoom Region Mode icon to zoom in on the font in the middle of the scene (see Figure 1–14).

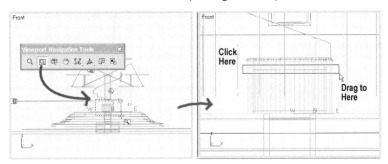

Figure 1–14 *Zoom in on the font in the Front viewport*

3. Right-click to cancel the zoom mode.

 TIP If you have a wheel mouse, you can roll your wheel instead to zoom in and out.

4. Move your cursor to an area where there is no geometry, and then click and hold down the mouse (see the right side of Figure 1–14).

5. With the mouse held down, drag it to create a selection window so that you include only the geometry (an AEC Polygon) used for the water in the font (it appears as a green line in this view).

6. Release the mouse to select it.

7. In the Modify panel, change the name (currently "A-Poly") to **Font Water**.

   This will make it easier to find and select later.

Near the bottom of the font, there is another AEC Polygon that we should rename as well.

8. Select it the same way as the water surface.

   Be careful not to select the compass with the A-Poly object. If you do, simply select again and make the window a bit higher.

9. On the Modify panel, change the name to **Font Bottom**.

10. On the Modify panel, click on UVW Map button.

We are doing this because the map we will use for the tile pattern needs to know how to map itself to this surface. Most of the other ADT entities (other than AEC Polygons) already have this mapping.

11. Locate the Material palette (its default position is the left side of the screen), and then click the Scene — In Use tab to make it active.

12. Right-click the MVIZr – Tile Blue material and choose: **Apply To Selected**.

Create a Water Material

Let's create a new material for the water in the font. Making good water is a challenge. Let's explore the parameters necessary to make a believable water material.

13. From the tool palette, click the "Scene – Unused" tab to make it active.

14. Right-click on the Create New tool, and then choose **New > Water** (see Figure 1–15).

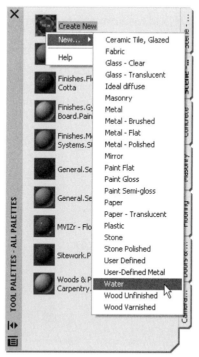

Figure 1–15 *Create a New Water material*

15. A new Water material will appear; right-click on Water and choose **Properties**.

16. Rename the Material from "Water" to: "**MVIZr - Font Water.**"

17. Expand the Texture Scaling rollout.

This is the important setting; it will affect the settings for the bump map that we will apply to make the surface of the water look believable. "Good water" is subjective. The following are suggested settings; feel free to adjust them as you see fit.

18. Input 1" for the Width and Height.

 NOTE VIZr defaults to feet, even if the ADT file is set to inches. Therefore, be sure to use the inch mark.

19. Expand the Special Effects rollout.

20. Click on the button labeled "None" to the right of Bump.

Buttons such as these are referred to as "slots." This would be considered the "Bump Slot."

21. In the Material/Map Browser, double-click on the **Waves** procedural map.

22. Rename the Map to: "**MVIZr – Font Water Bump**."

23. In the Waves Parameters rollout, set Num Wave Sets to: **50**.

 NOTE Unlike the values used in the Width and Height above, these fields accept only absolute values and are relative to the unit used in Texture Scaling for the procedural map, therefore do not input a unit symbol for this or the values in the next two steps, simply the number is sufficient here.

24. Set Wave Radius to: **120** and Wave Len Max to: **2**.

25. Set Wave Len Min to: **1** and Amplitude to: **1** (see Figure 1–16).

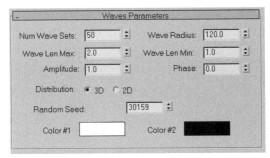

Figure 1–16 *Set the parameters of the Waves Procedural map*

Leave the remaining settings as is. These values have a direct relationship with the value you set under Texture Scaling, which in this case was set to one inch. Therefore, there will be 50 sets of waves with 120 inches between the first and last wave set. The maximum wave length is 2 inches, the minimum is 1 inch, and Amplitude is 1 inch. Basically, this means that each wave will vary from 1 to 2 inches in height. Had we not designated inches above, these values would be relative to the default unit in the scene or feet in this case.

26. Close the Material Editor.

Apply the Material

Now that we have our Water Material Defined, let's apply it to the AEC Polygon object in the font that we renamed above.

27. On the Selection toolbar, click the Selection Floater.

28. In the Selection Floater, double-click on the **Font Water** in the list and then Close the Selection Floater.

29. On the Materials palette on the Scene — Unused tab, right-click on Water and choose **Apply To Selected**.

Render the Water

Before we render again to see what our scene looks like after making these material changes, let's adjust the daylight system so that some sun hits the font to add a little more light on the water. This will help the water stand out better. Let's also reset the Radiosity before the test render as well.

30. Select the Daylight object from the Selection Floater.

31. In the Modify panel under the Control Parameters rollout, change the Month to 1.

32. From the Render menu, choose **Radiosity**.

33. Click on the Reset All button. Answer Yes in the dialog that appears.

34. Be sure that **Camera01** is chosen from the viewport list at the bottom (choose it if it is not), and then click the Render button at the bottom of the dialog (see Render Stats 01.04).

 Render Stats 01.04

Render Time: 13 seconds
Radiosity Processing Time: 4 seconds
Music Selection: "Rally" Prince of Egypt – Soundtrack (1998)
Color Plate: NA
JPG File: Rendering 01.04.jpg

REFINE RENDER SETTINGS

It's time to make a few adjustments to our rendering parameters to set up for a mid-range rendering.

Adjust Radiosity Settings

Let's begin these adjustments with a few tweaks to the Radiosity settings.

1. Select the Drape Mass Element (the hill upon which our model sits).

2. Right-click and choose **Rendering Properties**.

3. Beneath Geometric Object Rendering Properties, remove the checkmark from the Cast Shadows checkbox.

4. Beneath Geometric Object Radiosity Properties, place a checkmark in the Exclude from Regathering checkbox.

5. Deselect the Use Global Subdivision Settings and the Subdivide checkboxes.

6. Click OK to dismiss the Render Properties dialog.

This is done because we are going to be using a Radiosity solution. One of the aspects of the solution is that it will break down large faces into smaller triangulated areas so that varying light levels on the same surface can be represented properly. This is referred to as "Meshing." This process, while often dramatically improving the quality of a rendering, can also dramatically increase rendering times. Therefore, since this in an interior scene, meshing the huge landscaping area outside makes no sense.

7. From the Render menu, choose **Radiosity**.

8. Click the Reset All button, and then click Yes in the confirmation dialog that appears.

9. At the bottom of the dialog, choose the **Production with Radiosity** Preset from the Preset list.

10. In the Select Preset Categories dialog, be sure that all items are selected and then click the Load button.

    If a "Change Advanced Lighting Plugin" confirmation dialog appears, simply click Yes to accept and dismiss it ('Advanced Lighting Plugin' is simply another way to refer to 'Radiosity').

Modify Render Settings

11. On the Common tab, beneath the Photorealistic Render rollout, place a checkmark in the Antialias Geometry checkbox and set the Shadow/Material Quality to: **5**.

This will reduce the jagged look of diagonal lines in the final rendering.

12. Scroll down and expand the Render Options rollout. Place a checkmark in the Reflections/Refractions checkbox.

As you may guess, this allows us to see reflections in the glass and the refractions of the water.

13. In the Advanced Options area, set the Ray Bounces to: **2** (see Figure 1–17).

Ray Bounces is how many times the reflections and refractions are calculated. A setting of 1 will allow us to see the reflection of the circular wall in the window but not through the glass. A setting of 2 will give us both, as shown in Figure 1–17. The higher the number, the more accurate the glass and water materials will render. However, this will be at an exponential increase in rendering time.

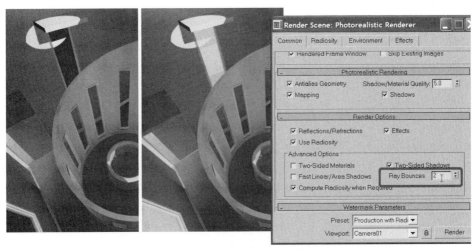

Figure 1–17 *See the effect of making adjustments to Ray Bounces*

In the image on the right you can now see through the glass. The reflections of the room are there even though the brightness of the exterior washes it out.

14. Switch to the Radiosity tab.

15. Under the Radiosity Processing parameters rollout, change the Initial Quality to: **85%** and the Refine Iterations (All Objects) to: **25**.

This will process the Radiosity solution to a much higher quality than the draft setting of 30%.

16. Expand the Radiosity Meshing Parameters rollout.

17. In the Global Subdivision Settings area, select Enabled and set the Meshing Size to: **5'-0"** (see Figure 1–18).

As mentioned above, the meshing will better approximate the lighting levels across larger surfaces. The dimension here represents the maximum size that any given triangle of the mesh will be.

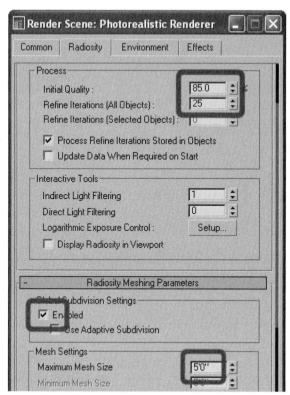

Figure 1–18 *Configure the settings for Radiosity*

Move to the Environment tab next.

18. On the Environment tab, in the Background rollout, click the color swatch. Set the colors to Red: **196**, Green: **210**, Blue: **244**, and then click Close (see Figure 1–19).

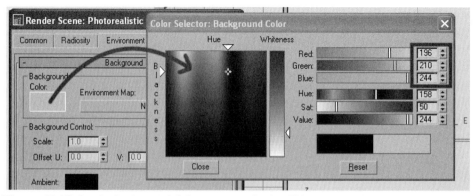

Figure 1–19 *Set the background Color to a nice shade of blue*

19. Scroll down the Environment tab, and in the Logarithmic Exposure Control Settings rollout, change the Brightness to **45**.

We need to up the brightness now because earlier when we lowered it, we did not have the surfaces meshed; so the average brightness across the surface was what the whole surface was being rendered at. Now, with the mesh, the bright parts can be bright and the dark parts can be dark.

20. At the bottom of the dialog, be sure that the Viewport is set to **Camera01**, and then click the Render button (see Render Stats 01.05).

 Render Stats 01.05

Render Time: 7 minutes 30 seconds
Radiosity Processing Time: 1 minute 41 seconds
Music Selection: "Blue Monday" New Order – Substance 1987 (Disc 1)
Color Plate: NA
JPG File: Rendering 01.05.jpg

HIGH QUALITY RENDERING

Rendering is largely a balancing act between quality and time. Often to achieve high quality output you must allow the rendering to process for significant amounts of time. It is a proportional relationship: higher quality requires longer render times, and vice versa. Therefore, you will often choose a compromise between a level of acceptable quality and time expended. High-quality renderings are often processed overnight to maximize computer resources and billable business hours. If you are happy with the results of the rendering produced just now, feel free to move on to Chapter 2 and the rest of the book. However, if you wish, you can load a provided Render Preset with higher quality settings and then render the scene again. This high quality rendering took approximately 1/2 hour on our test system. Your times will naturally vary depending on the specs of your hardware. While you may not need to run the rendering overnight, you might want to start the rendering and then take a lunch break.

1. Return to the Render dialog box.

2. At the bottom, load *Ch01_Final Render_Mid Range.rps* Render Preset.

3. Render the scene (see Render Stats 01.06).

 Render Stats 01.06

Render Time: 25 minutes 11 seconds
Radiosity Processing Time: 1 minute 44 seconds
Music Selection: The Cars – The Cars (1978) Entire CD
Color Plate: NA
JPG File: Rendering 01.06.jpg

Some minor changes to the Radiosity Processing Parameters were made as well as enabling the Regather Indirect Illumination setting. These settings will be covered in more detail in upcoming chapters.

In this release of ADT, a pretty extensive set of presets ship with the product. Feel free to explore them in place of or in addition to the ones we have provided here.

ADDITIONAL EXERCISES

Additional exercises have been provided in Appendix A. In Appendix A you will find instructions to load another Render Preset and create a high-quality rendering. This exercise will take some time to complete and is optional. Please view Color Plate C–1 to see what the final rendering will look like (see Render Stats 01.07).

 Render Stats 01.07

Render Time: 3 hours 30 minutes 47 seconds
Radiosity Processing Time: 5 minutes 54 seconds
Music Selection: Watch the "The Godfather" Part I (1972) on Widescreen DVD
Color Plate: C–I
JPG File: Rendering 01.07.jpg

CONCLUSION AND SUMMARY

In this quick tour of the features of VIZ Render, we have been exposed to many of its features and much of its potential. Among the features we explored, we learned the following:

Views in ADT become Cameras.

Cameras can be adjusted directly in the viewport with the tools provided in VIZ Render quite easily.

We can light our scene convincingly with the VIZ Render Daylight System.

Model geometry in ADT can be edited and then reloaded in VIZr.

Objects from the linked ADT Model can be selected in the Selection Floater dialog.

We can work with the Materials imported via the File Linking process from ADT or create our own.

Preparation for final high-quality rendering involves an iterative process of adjusting settings and then making test renderings to judge the results.

CHAPTER 2

User Interface and General Overview

INTRODUCTION

This chapter is designed to get you acquainted with the user interface and work environment of VIZ Render (VIZr) and to explain the basic requirements for its use. Feel free to experiment along the way as we explore the many features of the VIZr user interface and the ways that it interacts with ADT.

OBJECTIVES

The major goal of this chapter is get you acquainted with the user interface and basic tools available in VIZ Render. Much of the VIZr interface is different than ADT. Therefore, a major objective of this chapter is for you gain comfort with these differences and to also expose any similarities. In this chapter we will:

- Understand what VIZ Render is and how it relates to ADT
- Gain comfort with the user interface
- Explore the environment
- Work with Transforms and other Tools
- Understand Cloning, Merging, Rendering and Radiosity

THE 3D PRESENTATION PROCESS

To create 3D presentations on a computer, you follow a process similar to a photographer in his or her studio. As we stated at the start of the last chapter, we will divide our time into essentially three equally important areas: Modeling, Lighting/Materials, and Rendering. Just like the photographer, we must first have a subject for our photograph—this is the model. A photographer often spends a great deal of time getting the lighting conditions and other details of the scene just right. This is analogous to the Lighting/Materials stage of 3D presentation and then finally the photo is shot, processed, and potentially re-touched to yield the final result. All of these tasks are applicable in the creation of the 3D rendering. Before we explore the specific user interface and tools of VIZ Render, let's spend a little time on each of these items and define some working definitions.

MODELS

Architectural Modeling can be classified in a few different ways. Before computers, when an architect spoke of a model, they were usually referring to a scaled down version of the building design made from foam core, chipboard, or wood. This type of "physical" model certainly still exists in architecture, but today the word model can invoke many different meanings. Let us begin this chapter with a discussion of the various types of models that are prevalent among architects using computer design software such ADT and VIZ Render. For the purpose of this discussion, we will use the following working definition of the term *model*.

Model—a "scaled" representation of a "real" thing. Therefore, "dimensionality" (two- or three-dimensional) is not what makes a model, but rather that it is a *to scale* representation of an *actual* object or thing you can ultimately touch.

This working definition shares the "scaled" part of the original notion of model as mentioned above; however, this definition is broader in scope. By this reckoning, an architectural drawing would be considered a model. After all, a drawing is scaled, and does represent a "real" thing. Although this is true, typically a model in this context would be limited to the on-screen display, and the output of such a model is nearly always referred to as a drawing. With these caveats in mind, consider the following types of computer models in ADT and VIZr:

Two-dimensional Production Models—this is the most common type of model produced with CAD (Computer Aided Design) software. Nearly all drawings produced in core AutoCAD without ADT objects would be classified as 2D production models. As we have said, these models are typically referred to as "drawings" rather than models. The most prevalent use of such a model is for construction documentation. 2D Production Models offer little benefit in 3D rendering.

Three-dimensional Presentation Models—this is a model produced purely for the purpose of generating a 3D rendering. Elements of the design are only refined to the level of detail required for the rendering. They serve to convey the intent and feel of a design rather than a true representation of *all* that will be there when the project is complete. For instance, a 3D Presentation Model may include only the exterior shell of a building and then perhaps only those walls actually required to generate the view from which the actual rendering will be shot.

Three-dimensional Design Intent Models—this is the type of model typically produced by Architectural Desktop. A Design Intent Model typically contains a mix of 2D and 3D geometry. The level of detail in this sort of model operates best at overall building scales such as 1/4"=1'-0" [1:50] or 1/8"=1'-0" [1:100], and are meant to convey overall building design and intent. They therefore contain only major building components and rely on traditional 2D Production Models to convey detailed construction components and assemblies.

Three-dimensional Virtual Building Model—this type of model would include all building components accurately represented in 3D. This type of model is still very rare in Architecture, Engineering, and Construction (AEC). The goal of a model such as this is full coordination of building components and systems long before construction takes place.

Building Information Model (BIM)—in many ways BIM is a compilation of several of the other model types. A BIM shares the same goals as a 3D Virtual Building Model, namely the full coordination of building components and systems, without the requirement that all components in the model be conveyed in 3D geometry. Much of a BIM's intent can be conveyed through data properties assigned to the objects rather than literal modeling of all components. A successful BIM would balance the value of fully coordinated 3D geometry with the potential diminished returns of "over modeling" small-scale details. However, rather than eliminate those details, or disassociate them into accompanying 2D detail drawings as a Design Intent Model does, the BIM approach seeks to incorporate such details in the model using data properties and other such "non-graphical" data, and, most importantly, to maintain a dynamic relationship between all such components. For a model to truly be considered a "BIM," all components, whether 2D or 3D, graphical or non-graphical, must be fully coordinated into a unified whole.

Evolution of the Process (with ADT and VIZr)

Without Architectural Desktop, most firms generate a 2D Production Model for CDs and then a separate 3D Presentation Model for client presentations and rendering. Even though the 2D model in this type of process is often used to generate the 3D model, the obvious problems with this approach are the potential disconnect between the two models which could easily get out of sync as design changes occur, and duplication of effort, since two separate models are required. However, the advantage of this process is that each model contains only the data and geometry required to meet its intended purpose: there is no 3D data in the production model to confound the CD process, and all of the notes and dimensions and other CD data are left out of the 3D presentation model.

Although this approach is still valid and prevalent in many firms, with Architectural Desktop and VIZ Render the goal is to build a single model that is equally suited for both needs. When used properly, these tools allow us to attain this goal without the need to build two separate models. Users are encouraged to begin migrating toward the creation of one of the latter three model types defined above (Design Intent Model, Virtual Building Model, or Building Information Model). For most ADT users, initially this will mean a Design Intent Model that can be used for both CDs and Presentation Renderings. Often you will end up with a hybrid of one or more types of model depending on the goals and specific needs of your project.

In Chapters 3 and 4, we will construct a model that will be used to create high quality renderings in Chapters 5 and 7. The model that we will build is early in the schematic design phase of the project, and all that has been designed is the façade of the proposed townhouse. This could easily lead to the creation of a 3D Presentation Model simply for the purpose of winning over the client. Since we are only talking about a single façade, this approach could easily be justified even if the model thus generated were not transitioned to the CDs. However, by taking a bit of time to plan the organization of the file(s) for the project and using good ADT modeling techniques, we have much more potential to see this model transition smoothly to the next phase of the project and become a Design Intent Model. Naturally each project is unique, and though it is a useful and valid goal to build more sophisticated and robust models, there will certainly be times when conventional wisdom and project restrictions such as budget limits will lend themselves to the more traditional approach. Remember to think in terms of "evolution" rather than "revolution" in these endeavors, and take each project's special circumstances into account when you decide.

Linked Geometry

Modeling in VIZ Render is nearly nonexistent. With the exception of applying some types of modifiers to 2D geometry, all modeling is completed in ADT. Since VIZ Render relies on ADT for geometry, a file link mechanism is provided to import the geometry from an ADT file and maintain a link back to the source file. This file link is maintained so that changes made to the ADT model may be refreshed within VIZr. The process of linking to VIZr from an ADT file will create a new VIZr file in the same location and with the same name (but with a .DRF file extension) as the drawing file being linked. If a DRF file of this name already exists in that location, the existing file will be loaded instead.

LIGHTING AND MATERIALS

Lighting is extremely important in conveying the *feeling* of a space. Too much light in a space can wash out subtle features, whereas a space that is poorly lit can lead to a dark or claustrophobic feeling. Proper lighting of a scene is critical when conveying the architect's vision for a building or space. Creating good lighting is as much art as it is science. "Good" lighting is obviously very subjective, and the right amount and type of lighting required to convey a desired mood can vary considerably from one project or scene to the next.

Materials in VIZr are the graphical representation of real world building materials such as glass, concrete, or steel. These may be represented by using photographs of the actual product applied as a bitmap to the surfaces of the 3D model geometry, or VIZ Render can simulate many material characteristics using color and mathematically generated procedural maps. For example, in Chapter 1, the tile material applied to the font used an actual photograph of ceramic tile. Conversely, the water was a procedural "waves" map that implied the look and feel of actual water.

 NOTE While Material Definitions can be shared back and forth between ADT and VIZr, procedural maps are not supported and therefore will not display in ADT. For this reason, most of the out-of-the-box Material Definitions use bitmapped textures.

The success of your VIZ Render scenes will rely as much on how skillful you become at manipulating lighting and materials as your craftsmanship in constructing models.

RENDERING

The word "rendering" is used frequently in relation to 3D computer software. It often refers to the output of an image file from a 3D model. Among the many uses of the word "render" as defined in the *Merriam-Webster Dictionary* we find: "to reproduce or represent by artistic or verbal means: Depict." From this use of the word and this definition we can derive the function of the term *rendering* as it applies to VIZ Render (and other 3D graphics software packages). For the purposes of our discussions in this text, the following definitions will apply:

▶ **Rendering**—a 2D image generated from a 3D model in computer graphic software (specifically ADT and VIZ Render, in this case).

▶ **Render**—the action of generating a rendering in the software.

Just as achieving desired results in materials and lighting involves equal measure of art and science, the final result of your rendering relies heavily on your ability to skillfully master many intricate and sometimes disparate settings. A photographer may have the perfect subject, may have a carefully planned scene with expertly positioned and artfully placed lighting. But if the settings of the camera itself are not adjusted with care and knowledge, the final exposures can be dismal. Conversely, with all the above arranged just so, and with careful attention paid to the settings of the camera, the same scene can yield stunning photos worthy of an artist's gallery.

THE CHAPTER 2 DATASET

The dataset used in Chapter 2 is a simple model of a small room with a counter in the center. The camera vantage point has already been established for this scene. We are looking into the room made from two walls and a floor. At the moment, it is missing the other two walls and its ceiling. (We will add these later). The two walls that we are facing contain some simple windows. Finally, three light fixtures (made from Mass Elements) are suspended above the counter.

Start Autodesk Architectural Desktop and Load a File

1. If you have not already done so, install the dataset files located on the Mastering VIZ Render CD ROM.

   Refer to "Files Included on the CD ROM" in the Preface for information on installing the dataset files included on the CD.

2. Launch Autodesk Architectural Desktop from the icon on your desktop.

   You can also launch ADT by choosing the appropriate item (Autodesk Architectural Desktop) from your Windows Start menu in the Autodesk group. In Windows XP, look under "All Programs"; in Windows 2000, look under "Programs." Be sure to load Architectural Desktop—not VIZ Render— for this exercise.

3. From the File menu, choose **Open**.

4. Choose *My Computer* and then your C: drive.

5. Double-click on the *MasterVIZr* folder, and then the *Chapter02* folder.

6. Double-click on the *File Link.dwg* file to open it.

View the Drawing in 3D

This is an extremely simple dataset. However, it will provide a suitable backdrop to explore the next several topics. In Plan it is simply two Walls and an element in the center of the room.

7. From the View menu, choose **3D Views > SW Isometric View**.

8. From the View menu, choose **Shade > Hidden**.

As you can see, there is counter in the middle of the space and three light fixtures suspended above. Let's bring this file into VIZ Render and begin exploring that application's user interface.

9. From the Open Drawing Menu, choose **Link to VIZ Render** (see Figure 1–2 in Chapter 1).

THE VIZ RENDER USER INTERFACE (UI)

Now that we have defined some of the critical terminology that we will use throughout this book, let's begin exploring the VIZ Render interface and toolset. For the next few minutes, we will take the "whirlwind" tour of the VIZr tool set. All of the tools covered in the following steps use default VIZ Render settings. The chapters that follow will employ much of what you discover here.

The file that we have just linked will provide a suitable backdrop to explore the elements of the VIZr user interface. The VIZ Render user interface shares a few elements in common with most other Windows applications: the Menu Bar across the top edge to the left, Toolbars that can be either docked or floating, and a Status Bar along the bottom edge. Like ADT, which also has each of these elements, VIZr also has Tool Palettes. Unique to VIZr are the Time and Animation Controls and the Command Panel (see Figure 2–1 and Color Plate C–2).

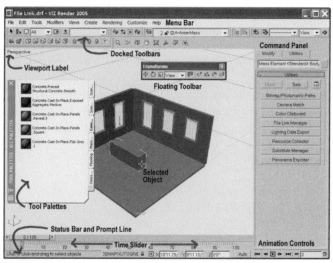

Figure 2–1 *The VIZ Render user interface*

STATUS BAR AND ANIMATION CONTROLS

The Status Bar is used to give feedback during certain commands and procedures. The Animation Controls and the Time Slider are used to create and play animations. The specific functions of the Status Bar and its many mode toggles will be covered both below and in later chapters. The Animation Controls and the Time Slider are covered in Chapter 9.

MENUS AND TOOLBARS

The menus and toolbars function as they would in any Windows application. Click a menu item to open it and choose a command. Hover over a toolbar icon to see a tooltip of its function. If you wish to load a toolbar that is not visible on-screen, right-click next to the docked toolbars in the blank gray area to open a menu of available toolbars (see the left side of Figure 2–2 and Color Plate C–2).

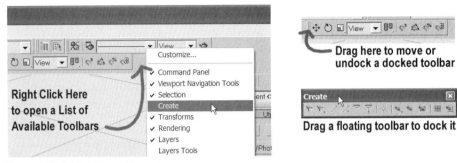

Figure 2–2 *Working with toolbars*

The toolbars currently loaded will appear in the list with a checkmark next to their name. Choose one that does not have a checkmark to load it. Choose one with a checkmark to hide it. You can dock your toolbars to the side of the screen by dragging them there. You can undock a toolbar in the same way. Drag the double gray vertical bar to the left of a docked toolbar to move it or undock it (see the right side of Figure 2–2 and Color Plate C–2).

QUAD MENUS

Like most Windows applications, when you right-click the mouse, a context-sensitive menu will appear. In VIZ Render, such a menu is called a "Quad Menu." It is called "Quad" since it has the potential for four separate menus organized in quadrants. The two menus on the right side contain commands common to all VIZr objects. The menus on the right vary depending on the object selected. Use of Quad menus puts all of the most commonly used commands and options for a particular object right at your fingertips (see Figure 2–3).

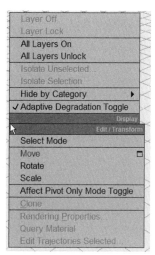
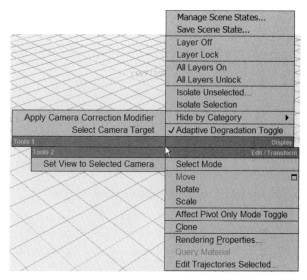

Quad Menu, No Object Selected   **Quad Menu with a Camera Object Selected**

Figure 2–3 *Typical Quad menus in VIZr—common commands on the right, object-specific on the left*

TOOL PALETTES

Tool Palettes in VIZ Render are used to create, edit, and manage a scene's materials. The Tool Palettes included with VIZ Render include two static palettes: Scene - In Use and Scene - Unused. There are also five additional sample palettes: Concrete, Masonry, Flooring, Doors & Windows, and Cameras and Lights. Click on a tab to switch to that palette. Let's explore some of the functionality of the Tool Palettes now.

 NOTE If you are already familiar with Tool Palettes in Architectural Desktop, the VIZ Render Tool Palettes behave in much the same way. If you wish, you can skip this topic.

New to VIZr 2006, you are able to create and organize Tool Palette Groups.

Tool Palettes can move to either side of the screen. You can use the small icon on the bottom of the title bar (or the option on the right-click menu) to set the Palettes to "auto-hide." With Auto-hide enabled, the Palettes will collapse to just the title bar when the mouse is not over them. When you move the mouse over the title bar, they pop open again. (see the left side of Figure 2–4 and Color Plate C–2).

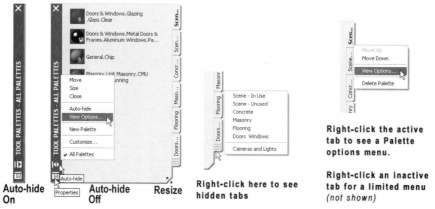

Figure 2–4 *Working with Tool Palette options*

Depending on the size of your Tool Palettes, some of the tabs may be hidden from view. This will be indicated by the small group of bunched up tabs at the bottom of the tabs. To see hidden tabs, click here to see a list of all available palettes (see the middle of Figure 2–4). If you wish to configure the properties of the Tool Palettes group, right-click the title bar, as shown in the left side of Figure 2–4. If you wish to configure the properties of a single palette, right-click on its tab, as indicated in the right side of Figure 2–4.

Tool Palettes can be displayed in one of three ways: List View, which shows an icon with the text label next to it; Icon with Text view, which shows the icon with a label beneath it; or in Icon only view, which shows just the icon without a label (see Figure 2–5).

Figure 2–5 *Tool Palettes can display in a variety of ways*

If you wish to change the View options, right-click the Tool Palettes and choose View Options. In the View Options dialog, choose your desired View Style. You can also change the size of the icon and optionally have your settings applied to all of the palettes rather than just the current one (see the left side of Figure 2–5).

VIEWPORTS

The VIZ Render work area is divided into one or more Viewports. Each viewport can be assigned to a particular vantage point such as an orthographic view like Top, Left, or Front, a Perspective view, or a Camera view. The current view is indicated in the top-left corner of each viewport by a viewport label (shown in Figure 2–1). The VIZr interface can be divided into one, two, three, or four viewports. Several combinations are possible. If you are familiar with "Tiled Viewports" in ADT, the concept is similar. The screen must be divided completely (or tiled) into one or more viewports. The default configuration is four equal viewports, two wide and two high. This may not be apparent onscreen since you are likely seeing only a single viewport at the moment. Regardless of the viewport configuration you choose, you can always "maximize" the current viewport to fill the screen. This is the state we are seeing currently onscreen.

Understanding Viewports

Let's take a look at the viewport layouts that are available.

1. On the Viewport Navigation Tools toolbar, click the Minimize Viewport icon (see Figure 2–6).

 TIP Recall from Chapter 1 that you can also press **ALT + W to toggle between Minimized and Maximized viewports.**

Figure 2–6 *Minimize the current viewport on the Viewport Navigation Tools toolbar*

Notice that the current Perspective Viewport shifts to the lower-right corner of a four viewport configuration. We now also have a Top view in the upper left, a Front view in the upper right, and a Left view in the lower left.

2. Being careful not to click on an object in the scene, click anywhere in the Top viewport.

Notice the way that the edge of the viewport highlights. This indicates that it is the active viewport.

3. Repeat this step in other viewports.

Let's take a look at the available viewport configurations.

4. From the Customize menu, choose **Viewport Configuration**.

5. Click the Layout tab (see Figure 2–7).

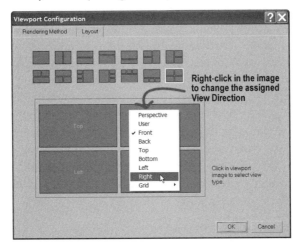

Figure 2–7 *You can choose or customize a viewport layout in the Viewport Customization dialog*

There are 14 configurations across the top. Simply choose one and then click OK to switch your display to that configuration. If you wish to edit the configuration, you can click on it at the top, then right-click each view below and choose the desired view direction for that view, as shown in Figure 2–7.

6. Choose any configuration at the top, edit its views if you wish, and then click OK to see the change.

   Repeat this process as many times as you like.

7. Place your mouse over the border between two or more of the viewports and drag (see Figure 2–8).

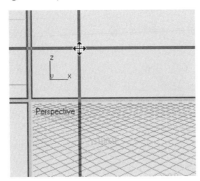

Figure 2–8 *Drag a viewport border if you wish to modify the proportions*

Notice that you can change the shape of the viewports in this way. You can customize any of the 14 configurations using these techniques.

8. Return to the default four equal configuration when you are done experimenting.

VIEWPORT NAVIGATION

Navigating a scene will naturally be very important; particularly when you have complex models. As you use the tools provided, you will become more comfortable with them. Several tools are provided on the Viewport Navigation Tools toolbar.

Zoom Extents

Let's begin with Zoom Extents. There are two such tools: Zoom Extents and Zoom Extents All. Zoom Extents will fill the current viewport with the entire model, while Zoom Extents All will do this for all viewports. In other words, VIZr will find the best zoom that will fit the entire model onscreen and fill the viewport with it.

If you did not Minimize the Viewports in the previous sequence, do so now. Click the Minimize Viewport icon shown in Figure 2–6, or press ALT + W.

1. Click your mouse in the Top viewport to activate it.

2. On the Viewport Navigation Toolbar, click the Zoom Extents icon (see Figure 2–9).

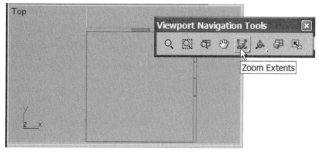

Figure 2–9 *Zoom Extents in the Top Viewport*

Notice that the Top view of the model now fills the Top viewport. This is always a good way to start before performing other zooms and pans. If you would like to zoom all viewports to extents, you can use the Zoom Extents All icon to do them all at once.

3. On the Viewport Navigation Tools toolbar, click the Zoom Extents All icon (see Figure 2–10).

Notice that all views now fill their respective viewports.

NOTE If you have a Camera Viewport active (see below), Zoom Extents has no effect.

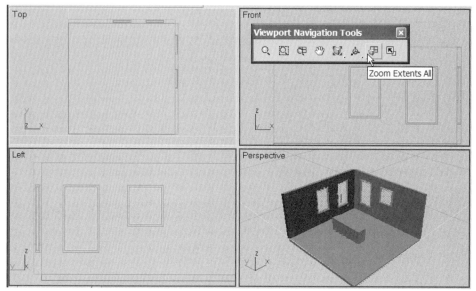

Figure 2–10 *Zoom Extents All to zoom all viewports*

Using the Wheel Mouse

If you have a wheel mouse, then you can Zoom the scene by rolling your wheel, and you can Pan the scene by pressing and holding down the wheel and then dragging the mouse. Release it to stop panning.

 NOTE If you don't have a wheel mouse, you may want to consider purchasing one. There are several inexpensive models available, and having access to the wheel will provide an immediate boost in productivity.

4. In the active viewport, roll the wheel on your mouse up a few clicks, and then down a few clicks.

Depending on the speed of your system, this should give a nice smooth zoom action.

5. Hold down the wheel and drag the mouse around.

6. Move your mouse over a different viewport and then press the wheel.

   This will set that viewport active and then you can roll or drag the wheel again.

If you are in a Perspective viewport, the motion will pan the perspective, and the model will appear to "fall off" from side to side. Panning in an orthographic view does not exhibit this behavior.

Zoom and Pan Icons

While zooming and panning with the wheel is very useful, in some cases, using the toolbar icon can give you greater control when attempting slight changes.

> 7. Click the Zoom Mode icon on the Viewport Navigation Tools toolbar.
>
> Holding down the left mouse button, drag in any viewport to zoom.
>
> 8. Click the Pan Mode icon on the Viewport Navigation Tools toolbar (see Figure 2–11).
>
> Holding down the left mouse button, drag in any viewport to pan.

Figure 2–11 *Zoom and Pan Mode icons*

Notice that you can get more subtle zooms and pans using the icons. This is useful when you wish to zoom or pan just a bit and the wheel is not quite giving you the control you need. Also, if you don't have a wheel, then these are the tools for you.

> 9. To cancel the Zoom or Pan modes, click the right mouse button.

Notice that you can switch from Zoom Mode to Pan Mode by simply clicking the respective icons. However, if you wish to stop zooming and panning altogether, then right-click the mouse.

Other Zoom Modes

There are a couple of other Zoom tools on the Viewport Navigation Tools toolbar.

> 10. Click the Zoom All icon (third from the left).
>
> 11. Drag in any viewport.

Notice that all the viewports zoom in and out as you drag the mouse. Right-click the mouse to cancel the command when you are finished. Zoom Extents All again after you are done.

> 12. Click on the counter object in any viewport to select it.

13. On the Viewport Navigation Tools toolbar, click and hold down the **Zoom Extents** icon (this opens a flyout icon), drag down, and choose the **Zoom Extents Selected** icon (see Figure 2–12).

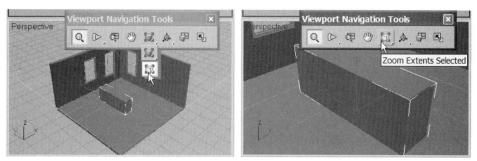

Figure 2–12 *Zoom Extents Selected zooms to the extents of the selected object(s)*

Notice that the selected object now fills the active viewport. You can reverse any zoom or pan action on the View menu.

14. From the View menu, choose **Undo View Change** (or press SHIFT + U).

You can step back through several previous view changes this way. Simply repeat the command.

SWITCHING THE ACTIVE VIEWPORT

We have been testing several navigation commands. It bears mention that you can switch active viewports with any mouse button. If you click to switch viewports with the left button, you will also be potentially selecting objects. If there is an object beneath your mouse when you click in the new viewport, it will be selected (and any existing selection will be deselected) and the new viewport will be made active. If you wish only to change the active viewport, use your right mouse button or the wheel of your mouse instead. Try it now with all three buttons to see for yourself. Be sure that an object is selected before you begin experimenting.

There is one additional way that you can maintain your current selection as you switch viewports, even with the left mouse button—use the Selection Lock Toggle on the Status Bar (see Figure 2–13).

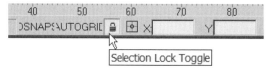

Figure 2–13 *Use the Selection Lock toggle to maintain the selection even as you left-click in another viewport*

Arc Rotate

With Arc Rotate, you can change the viewing angle in two or three axes simultaneously and interactively. Use this tool when you want to give your model a "spin" in the viewport.

15. On the Viewport Navigation Tools toolbar click the Arc Rotate icon.

A yellow circle with four handles at its top, bottom, left, and right will appear. If you drag the top or bottom handles, motion will constrain to the X-axis. The left and right handles constrain to the Y-axis, and outside the circle constrains to the Z-axis. If you drag in the center, you can rotate the view in all directions freely. If you use this tool in a perspective view, the top and bottom handles change the height of the perspective (moving your vantage point up or down). The left and right ones will rotate the ground plane. Rotating outside the circle in a perspective view is not very useful (see Figure 2–14).

16. Try the arc rotate tool in each viewport.

 TIP Remember that you can right-click before you release the mouse to cancel the rotation before completing it.

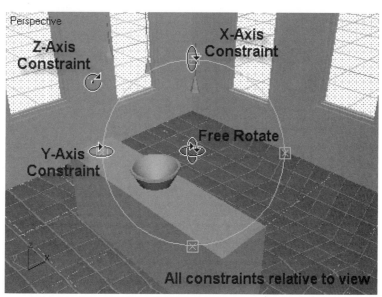

Figure 2–14 *Arc Rotate has various behaviors depending on where you drag*

TIP Remember that you can always use the Undo View Change command on the View menu if you want to reset a view change.

The last tool to look at is the Zoom Region. This tool changes to Field of View in the Perspective viewport. You can use it to zoom in on a particular region of the view by dragging a box around the area you wish to enlarge. After you are done experimenting with the various viewing tools, reset all of the viewports to their default viewpoints and then Zoom Extents All (see Figure 2–15).

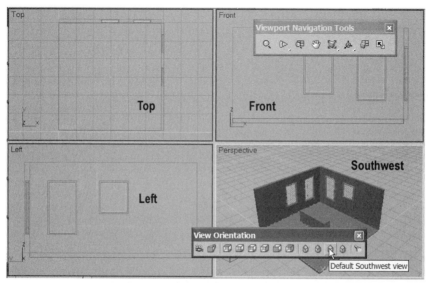

Figure 2–15 *Reset all Views to the default orientations and zoom extents*

You can use the View Orientation toolbar for this, or right-click on the Viewport Labels and choose the view from the menu (see the next topic).

VIEWPORT MENUS

In the upper-left corner of each viewport is a label indicating the viewing direction or the Camera name assigned to the viewport. If you right-click on this label, a menu will appear with a list of options for that viewport. Use the commands on this menu to change the current view direction in the viewport, change the active shading mode, and configure the viewport configuration (see below). Right-click the viewport label in one of your viewports to see the options. If you change any settings, be sure to reset them before continuing (see Figure 2–16).

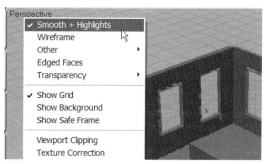

Figure 2–16 *Right-click a viewport label to access the Viewport Menu*

THE COMMAND PANEL

The Command Panel is similar to the Property Palette in ADT. The Command Panel is interactive and its parameters adjust based on what you have selected in the scene. This is where you access many of an object's parameters. We will be accessing the Command Panel quite a lot in this chapter as well as throughout the rest of the book. When there are no objects selected in the scene, the Modify and Utilities Command Panels will appear as shown in Figure 2–17.

Figure 2–17 *The Modify and Utilities Panels with no objects selected*

View an Object's Parameters

To see the editable parameters of an object, simply click on it and then look at the Modify Panel.

Be sure that the perspective viewport is current and Maximized.

1. Click on the box in the middle of the scene.

Notice that the selected object's name appears at the top of the Modify Panel and its parameters appear below. For any object selected in this simple scene, the object type is "Linked Geometry" and the only parameter is the "Reset Position" button. This button is used to return an object to its original position in the linked ADT file. Use this when you move or otherwise transform a linked object and then wish it to return to its original position (see Figure 2–18).

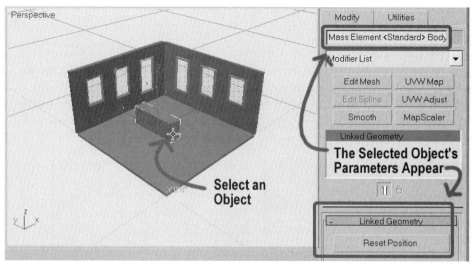

Figure 2–18 *Selecting an object reveals its name and parameters on the Modify panel*

Transforms

The three actions of Move, Rotate, and Scale are referred to in VIZ Render as Transforms. In other words, you can transform an object's location, orientation, or size. These three operations are combined with the action of selecting. Therefore, you can select and move, select and rotate, or select and scale an object. Use the Transforms toolbar, or the right-click "Quad" menus, to switch modes.

2. Load the Transforms toolbar (as indicated in the "Menus and Toolbars" topic above) and then click the Move icon.

 TIP You can also right-click and choose Move from the Edit/Transform Quad menu.

Notice the brightly colored red, green, and blue X-, Y-, and Z-axis tripod that appears in the center of the object. This is referred to as the "Transform Gizmo"—

seriously, look it up. You can use the Transform Gizmo to move an object and keep it constrained to one or two axes as you do (see Figure 2–11).

   3. Move your mouse over each axis, then over each of the yellow squares between each pair of axes (see Figure 2–19).

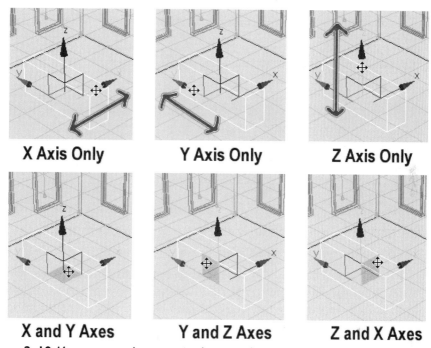

Figure 2–19 *Movement can be constrained to a single axis or a pair of axes using the Transform Gizmo*

Notice that they highlight to indicate the various possible constraints. If you click and drag the X-axis of the Transform Gizmo, for instance, the object will move only along the X-axis, while clicking and dragging the yellow square between the Z- and X-axes will move the object while constrained to the ZX plane.

   4. Highlight the X-axis of the Transform Gizmo. Click and drag the mouse without releasing the button.

Notice that the selected object moves only along the X-axis no matter which way you move the mouse.

   5. While still holding the left button, click the right button.

This will cancel the move operation returning the box to its original position.

     If you accidentally let go of the left button before canceling with the right button, choose **Undo Move** from the Edit menu and then try again.

6. Repeat the process for the other axes until you get the hang of it.

7. On the Transform Toolbar, click the Select and Rotate icon.

The Transform Gizmo will appear very similar, but the cursor will appear as a rotate icon when the mouse highlights an axis this time.

8. Try rotating along any axis. Again, use the right-click method to cancel if you wish.

9. On the Transform Toolbar (or the Quad menu), choose Scale and experiment with the various options (see Figure 2–20).

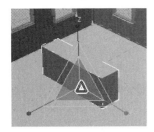 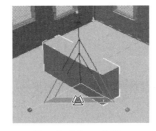 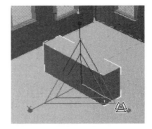

Scale X, Y and Z   **Scale X and Y**   **Scale only Y**

Figure 2–20 *Object selections can be scaled in one, two, or all three axes*

Regardless of whether you have used the right-click option to cancel a transform, or chosen Undo from the File menu to reverse it, you can always return linked geometry such as the object that we have selected here back to its original location, orientation, and size.

10. On the Modify Panel, click the Reset Position button (see the bottom-right corner of Figure 2–18).

11. Save the scene.

WORKING WITH FILE LINK

Now that we have explored the most common elements of the VIZr User Interface, let's take a closer look at File Link. As you recall from above, and the tutorial in Chapter 1, we use the file link process to link an ADT model to a VIZr scene. This is often accomplished with a simple one-step process as seen in the final step of the "View the Drawing in 3D" topic, above.

File Link Manager

1. From the File menu, choose **File Link Manager**.

Notice that the Utilities Command Panel automatically appears with the File Link Manager button depressed. As an alternative to the File menu, you can access the File Link Manager by simply clicking this button on the Utilities Panel.

Note the linked drawing file and its associated path (see Figure 2–21).

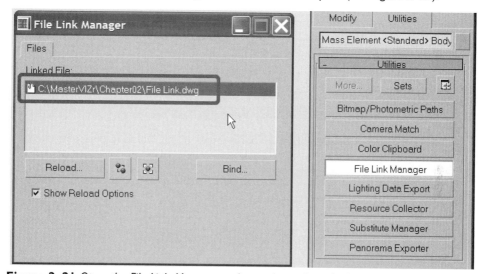

Figure 2–21 *Open the File Link Manager and note the path to the linked file*

Beneath this path are the Reload controls which are discussed below. If you prefer, you can also access the Reload options from the File Link toolbar.

2. Close the File Link Manager dialog.

RELOAD FROM ADT

When you execute the file link from ADT, several different behaviors are possible:

▶ *If VIZ Render is not running on your machine when you execute the File Link from ADT, a new VIZ Render session will launch and the ADT file will automatically be linked. A new VIZ Render scene (DRF file) with the same name as the ADT file is created in the same folder for this link.*

▶ *If the file that you are linking from ADT has previously been linked to a VIZr scene, VIZ Render will instead load this existing DRF file and automatically launch the File Link Manager prompting you to reload the file.*

▶ *If you have a session of VIZ Render open already when you perform the File Link from ADT, and the scene loaded in VIZr is the same as the file being linked from ADT, VIZ Render will launch the File Link Manager prompting you to reload the file.*

▶ *If VIZ Render is already running, but the scene does not match the current drawing in ADT, another session of VIZ Render will open with the file link. You can* therefore end up with several copies of VIZ Render running simultaneously, which can pose a big drain on computer resources. It is recommended that you avoid this practice.

Let's take a look at some of the many issues covered here. We will make some changes to the model back in ADT and then look at some of our options for re-loading that geometry in VIZr. In the upcoming steps a session of VIZ Render is open and the matching VIZr scene is loaded. Therefore, VIZ Render gains focus and we will be greeted with the File Link Manager dialog.

Modify the ADT Model

1. Make ADT active. (Click it on your Windows Task bar, or press ALT + TAB).

   ADT should still be running in the background. If you closed it, please re-open it now and then open the *File Link.dwg* file.

2. On the Walls Tool Palette, right-click on the CMU-8 tool and choose **Apply Tool Properties to > Wall** (see Figure 2–22).

   If the Walls Tool Palette is not visible, right-click the Tool Palette title bar and choose **Design** to load the Design Tool Palette Group.

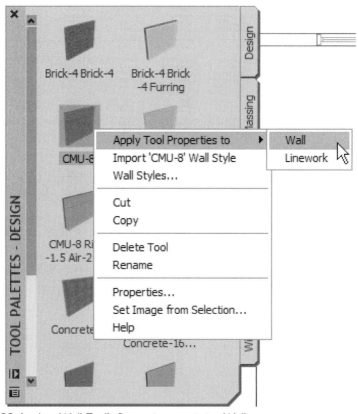

Figure 2–22 *Apply a Wall Tool's Properties to existing Walls*

3. At the "Select Wall(s)" prompt, select the left Wall and then press ENTER.

4. Click on one of the Windows, drag one of the height grips (the triangular shaped ones), and snap it to one of the gray bars (see Figure 2–23).

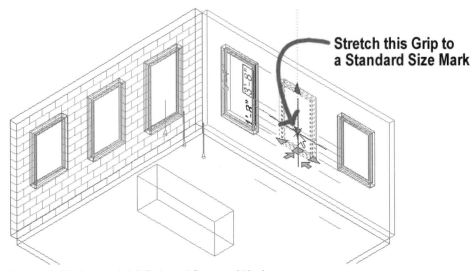

Stretch this Grip to a Standard Size Mark

Figure 2–23 *Swap a Wall Style and Resize a Window*

The gray bars indicate a "Standard Size" built into the Window Style.

5. Save the ADT file.

6. From the Open Drawing Menu, choose **Link to VIZ Render** (see Figure 1–2 in Chapter 1).

    VIZr will become the active window, and the File Link Settings dialog will appear.

7. In the File Link Settings dialog, accept the defaults and press OK.

Note that the left Wall now displays the CMU material that was assigned by the style in ADT. We will explore Materials in ADT more in Chapters 4 and 5.

RELOAD FROM VIZR

The ADT model can also be reloaded from within VIZ Render. There are two ways to do it. The first method is from the File Link Manager, shown in Figure 2–21 above. The second is from the File Link tool bar, as shown in Figure 2–30 and its accompanying descriptions below. Let's try this out now, but before we do, it will be easier to understand the various reload options if we first make a few edits to the VIZr scene.

Modify a Material in VIZr

1. From the Scene - In Use palette, right-click on the material named: Masonry.Unit Masonry.CMU.Stretcher.Running and choose **Properties** (see Figure 2–24).

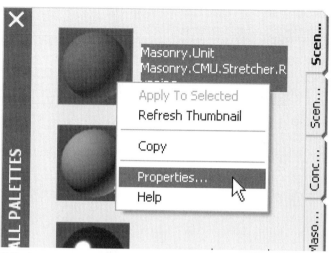

Figure 2–24 *Edit the properties of the CMU Material*

2. Click on the gray color swatch next to Diffuse Color.

3. Select a bluish color and then close the Color Selector dialog (see Figure 2–25).

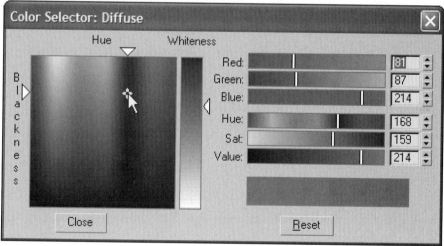

Figure 2–25 *Select a shade of Blue*

4. Next to the Diffuse Map entry on the right side, click the large button (labeled with the name of the material—see number 1 in Figure 2–26).

5. On the next screen, click the button labeled "Bitmap" just below and to the right of the preview panes (see number 2 in Figure 2–26).

6. In the Material Map browser, choose **NONE** and then click OK (see number 3 in Figure 2–26).

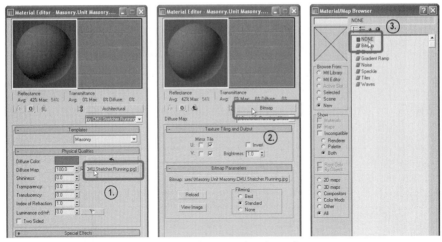

Figure 2–26 *Remove the bitmap and replace it with color only*

7. From the Scene - In Use palette, drag the Masonry.Unit Masonry.CMU.
Stretcher.Running material that we just modified and drop it on the right-
hand wall (see Figure 2–27).

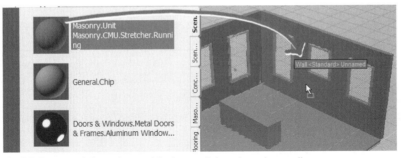

Figure 2–27 *Drag and drop the modified material to the other wall*

We have modified the masonry material so that it no longer resembles the original material assigned in ADT. We have also replaced the material that was applied to the right-hand wall with this same modified VIZr material. As we perform the next several steps, it is important to remember that the File Link between ADT and VIZr is *one-way*: from ADT to VIZr, and not back again. Therefore, both of these changes occur only in VIZr. Neither will be pushed back to ADT. At issue now is what to do with these types of changes when the geometry of the ADT model changes and you wish to reload the model. As we saw above in the "Reload Options" topic, there are several options for how to deal with materials changes in the VIZr scene when you reload the underlying ADT model. Let's explore those now. First, to make this test a bit more realistic, we should make a change to the geometry of the ADT model.

8. Click the Flooring tab to make this Tool Palette active.

This is a sample Tool Palette that is provided out-of-the-box with VIZ Render. It contains many sample materials that you can use right away in your scenes. Let's add a material from this palette to the floor surface now.

9. Drag Finishes.Flooring.Tile.Square.Terra Cotta and drop it on the floor surface. When you see a tooltip reading: Space <Standard> Floor, release the mouse button (see Figure 2–28).

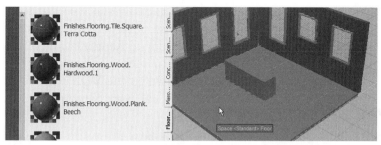

Figure 2–28 *Apply a material from one of the standard Tool Palettes to the floor*

If your floor tiles are "wavy," right-click the viewport label to open the viewport menu and choose **Texture Correction**.

10. From the File menu choose **Save**.

 NOTE It is important that you save here as we will be coming back to this saved state in the steps that follow.

11. Return to ADT, and delete the leftmost and rightmost Windows (see Figure 2–29).

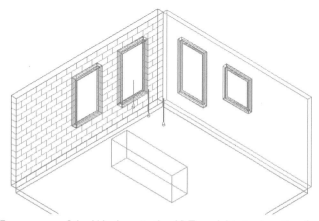

Figure 2–29 *Erase some of the Windows in the ADT model in preparation for File Link Reload Tests*

There is no need to save the ADT file as the File Link process is dynamic.

Do not choose the Link to VIZ Render option from the Open Drawing Menu (as we did above) this time. If you do this, you will not get the control to determine what happens to the VIZr material assignments. To gain this control, you must initiate the reload process from within VIZ Render.

RELOAD OPTIONS

As we have discussed and seen so far, geometry in a VIZ Render scene comes from a linked ADT model. The file link concept is much like an External Reference (XREF) within ADT. Just as an XREF can be reloaded any time to retrieve the latest changes made to the XREF file, a file link in VIZ Render can be reloaded at any time to retrieve changes made to the ADT model. Unlike an ADT XREF however, a file link will *not* automatically reload when the VIZr scene is loaded. All file link updates must be manually executed. There are four options to do so (see Figure 2–30).

Figure 2–30 *Exploring the various combinations of Reload Geometry and Materials*

Reload Geometry and Materials—Use this icon to reload an ADT model in VIZ Render. You can find the same function on the File menu named File Link Manager (see the leftmost panel of Figure 2–30).

When you reload an ADT model, there are two toggles that control the behavior of the materials imported from the ADT model. The first time you execute a file link, like we did in the steps above, all materials assigned to the ADT objects will automatically be imported and remain assigned in VIZr. In some cases, you will have defined new or edited existing materials in VIZ Render. You may also have changed the material assignments of linked objects in the scene. To preserve these types of changes upon reload, you can use the two toggle icons next to the Reload Geometry and Materials icon (or the equivalent icons in the File Link Manager dialog).

Use scene material definitions—with this toggle active (push the icon in), this option will maintain, in the scene, any materials you have edited in VIZ Render in favor of those being reloaded from Architectural Desktop. If this toggle is not enabled, the original material definitions from the ADT model will overwrite the ones in your VIZr scene. For example, let's say that you edit a brick material to use a different type of brick coursing. When the ADT model is reloaded, the ADT version of the material you edited still contains the old coursing. With this option enabled, you will maintain your edited coursing in VIZr. With it toggled off, the

ADT coursing will re-import and replace your edits in VIZr (see the second panel from the left of Figure 2–30).

Use scene material assignments (on Reload)—with this toggle active, this option will maintain, in the scene, any material assignments that you have made in VIZ Render in favor of those being reloaded from Architectural Desktop. For example, let's say that your ADT model contains a stud wall. In VIZr you assign a brick material to this wall. With this option enabled, this wall will remain brick even after the reload. With this option toggled off, it would return to being a stud wall (see the second panel from the right of Figure 2–30).

Use both Options—You can toggle both of the previous options on at the same time and preserve both material edits and assignments as geometry is reloaded (see the rightmost panel of Figure 2–30).

Bind (break link to drawing)—The final icon on the File Link toolbar is the Bind icon. This option will completely break the link to the ADT model file. This action cannot be undone. This is much like the XREF Bind option in ADT.

Updating the VIZr Scene

Again, as we saw in the "Reload Options" topic above, there are four reload options from within VIZr.

    1. Return to VIZ Render.

> **TIP** You can quickly switch between running applications in Windows by holding down the **ALT** key and then pressing **TAB**. **This will allow you to cycle between all running applications.**

    Be sure that the File Link toolbar is loaded. If it is not, refer to the "Menus and Toolbars" topic above for information on how to load it.

    2. On the File Link toolbar, click the "Reload Geometry and Materials" icon (see the leftmost panel of Figure 2–30).

Notice that the geometry of the ADT model has been updated: specifically the two Windows have been deleted. More importantly, you will notice that the material changes in the VIZr scene have been overwritten by their original designations in the ADT model. See the definition of "Reload Geometry and Materials" in the "Reload Options" topic above for more information on this behavior. Let's see what became of the materials that we assigned and edited in VIZr.

    3. Click on the Scene - In Use tab of the Tool Palettes.

Notice that the Terra Cotta material is no longer on the list.

    4. Click on the Scene - Unused tab.

Notice that the Terra Cotta material has been placed here. The Scene - Unused tab contains any materials that are in the VIZr scene file, but not applied to any objects. The materials that are not contained in the ADT model at all will be purged when you save and re-open the VIZ Render scene.

Before we move on and while we have the Tool Palettes in focus, let's take a quick look at something new in 2006, the Cameras and Lights palette

5. Click on the Cameras and Lights Palette.

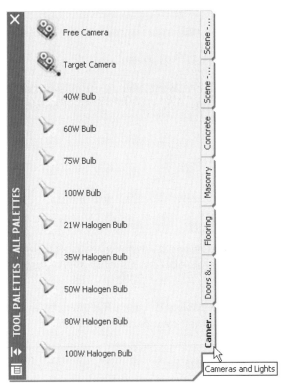

Figure 2–31 *The Cameras and Lights palette*

These tools place specific objects in your scene. They perform the same operation such as **Create > Photometric Lights > Preset Lights > 60W Bulb,** but with fewer steps.

Restore the Original Scene

Let's look at some of the other reload options now. To do this, we will recall the saved version of the file without saving the one we have onscreen now.

6. From the File menu, choose *C:\MasterVIZr\Chapter02\File Link.drf* in the recent files list.

    It should be the first file in the list.

NOTE Depending on your Windows file settings, you may see only the file name without the full path or file extension (.drf). This is display preferences only and will not impact anything here.

7. In the dialog prompting you to save the current scene, click the No button.

NOTE It is more likely that in most cases you would click "Yes" when prompted to save. However in this test, we are clicking "No" simply as a means to explore the various Reload options. If you click Yes in this case, you will actually not achieve the desired results in the next several steps.

8. On the File Link toolbar, click the "Use scene material definitions" toggle icon (depicted in second panel from the left of Figure 2–30) to enable this setting.

9. On the File Link toolbar click the "Reload Geometry and Materials" icon.

Notice that the floor and the right-hand wall have been reset to their state from the original ADT model. However, the wall on the left has retained its blue material. Utilizing the "Use scene material definitions" toggle preserved the edited version of the Masonry.Unit Masonry.CMU.Stretcher.Running material in the VIZr scene, but did not preserve any overridden material assignments. In other words, all geometry was reloaded and all material assignments were reloaded as well. However, the material definitions in the VIZr scene were preserved, allowing the CMU material to maintain its new settings.

10. Repeat the process above to restore the *C:\MasterVIZr\Chapter02\File Link.drf* in the recent files list of the File menu.

    Remember to choose No in the Save File confirmation dialog.

11. Click the "Use Scene material assignments (on reload)" icon to enable it (depicted in the second panel from the right of Figure 2–30), and then click the "Use scene material definitions" toggle icon again to turn it off.

12. On the File Link toolbar, click the "Reload Geometry and Materials" icon.

Note that this time the material assigned in ADT has regained its original settings (the CMU block texture reappears and the blue color is gone) and that VIZr retained the assignment of that material to the right-hand wall. Note also that the floor has maintained its tile assignment as well.

13. Restore the original file one last time without saving this version.

14. Toggle on both the "Use Scene material definitions" and the "Reload geometry and materials" icons this time (depicted in the rightmost panel of Figure 2–30).

15. On the File Link toolbar, click the "Reload Geometry and Materials" icon.

As you might expect, this time both material assignments and material definition edits have been preserved as the updated geometry reloads.

 IMPORTANT *Unless noted otherwise, when reloads are performed in this book, both toggles—"Use Scene material definitions" and "Reload geometry and materials"— will be enabled.*

We will keep this version of the *File Link* scene.

16. From the File menu, choose **Save**.

MERGE

The Merge tool allows you to import an entire VIZr scene or individually selected items from a scene such as geometry or a daylight system. Merge also supports 3ds max and Autodesk VIZ files. This allows you to mine the Web for content to help you populate your scenes. VIZ and max files have a *.MAX* extension. While you cannot simply open these files, if you purchase or are given such a file, you can use the Merge tool to import them into a VIZr DRF scene. Once these files have been merged into the current scene, they become part of the file. Merge does not create a link like ADT does. Merge is like inserting a block in ADT. Once it is merged, it is like any other object in the scene. Reloading the model from ADT will have no affect on merged items. They will remain part of the VIZr scene. However, they will *not* appear in ADT. Remember that File Link is one-way: from ADT to VIZr, not back.

Merge a DRF

17. From the File menu, choose **Merge**.

18. In the Merge File dialog, browse to the *MasterVIZr\Chapter02* folder on your C: drive.

19. Select *Bowl.drf* and then click the Open button.

A Merge dialog will appear that lists all of the objects contained in the file (only two in this case).

20. In the *Merge - Bowl.drf*, dialog click the All button at the bottom left of the dialog and then click OK (see Figure 2–32).

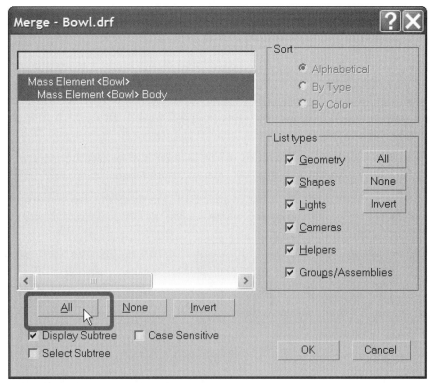

Figure 2–32 *Click All to select all objects in the list*

You will notice that the bowl is merged and is located at the origin of the existing scene. Merge automatically inserts the merged geometry at the origin of the current scene. You are free to move it once it is merged.

TIP If you know that you want the entire contents of the file to be merged, you can drag the file into the current scene from Windows Explorer. You will be prompted with a context menu where you can choose to either open or merge the file. The added benefit to this is that you will be able to locate the file wherever you like on the current XY plane. The origin of the merged file will be attached to the end of your cursor. Click to place it into the scene.

We will see many additional examples of Merge in upcoming chapters.

TOOLS MENU

Several very useful items are located on the Tools menu. These tools allow us to manipulate lights, cameras, merged geometry, etc. In the upcoming chapters, we will use several of these tools within our projects. For now, let's familiarize ourselves with some of the particularly useful ones.

Align

The Align tool allows you to match up a selection of geometry with some other target object. You can achieve a wide variety of alignments, such as the top face of this object with the bottom face of that one, or line up all right edges, and so on. For instance, our bowl has been merged into the current scene in the previous sequence. However, it is sitting on the floor. Suppose we wanted it on top of the counter. Using the Align tool, this is a snap.

1. Be sure that the bowl is still selected (it will have a white cage around it; if not, select it now). From the Tools menu, choose **Align**.

   Look at the bottom-left corner of your screen at the Status Bar and Prompt Line (see Figure 2–1 above). A message reading "Pick Align Target Object" will appear (see Figure 2–33).

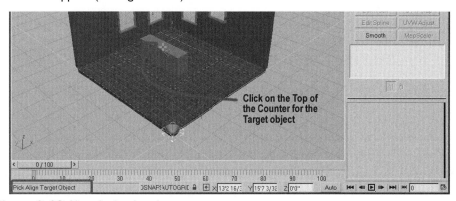

Figure 2–33 *Align the bowl to the counter top*

   Move the mouse around the screen. When it is over an eligible object, a cross cursor will appear; otherwise, only the Align symbol will show.

2. Move the cursor over the counter and then click on the top (see Figure 2–33).

The Align Selection dialog will appear showing several options. This dialog is interactive. This means that you can perform several alignments successively in a single session of this command. Simply use the Apply button in between each alignment.

3. Move the Align Selection dialog so that the bowl and counter are visible beyond.

4. Place a checkmark in the **X Position** (note the move of the bowl in the scene) and the **Y Position** boxes.

5. Click the Apply button.

This will center the bowl on the counter. Notice that the bowl is still selected, the Align Selection dialog is still active, but all checkboxes and other selections within it have been reset. The next task is to put the bowl on top of the counter.

6. Place a checkmark in the **Z Position** checkbox.

Notice the bowl move to the middle of the counter. This is because the Current and Target objects are both still set to the default of Center.

7. Choose **Minimum** for Current Object and choose **Maximum** for Target Object (see Figure 2–34).

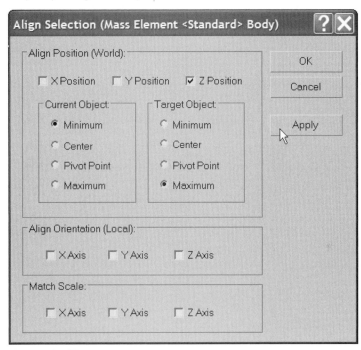

Figure 2–34 *Align the Z position*

This aligns the bottommost edge (minimum) of the selected object (the bowl) to the topmost surface (maximum) of the target object (the counter). Let's get a little more practice with this Align. To do so, we will add a light source to one of the spots over the counter.

8. Click OK to close the Align Selection dialog.

9. Save the scene.

Add a Light

At the moment, these spots are just modeled geometry and do not shed any light to the scene. By adding a VIZ Render light object, we can begin to illuminate this scene. The Align tool will help us get the VIZr light object positioned relative to the geometry that represents the light fixture. By positioning the light properly, the light it casts in the scene will appear more realistic.

10. From the Create menu, choose **Photometric Lights > Preset Lights > Halogen Spotlight**.

11. Click anywhere in your scene to place the light (see Figure 2–35).

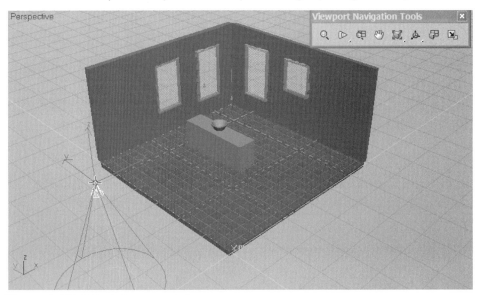

Figure 2–35 *Add a halogen spot light*

12. From the Tools menu, choose **Align**.

13. At the "Pick Align Target Object" prompt, click on the cone suspended above the counter that is closest to us.

You can verify that you selected it by reading the title bar of the Align dialog. It should read something like: "Align Selection (Mass Element <Standard> Body)."

 TIP If you have trouble selecting the cone, you can zoom in closer using the wheel on your mouse. Simply roll it up a few clicks in the viewport. If you don't have a wheel mouse, try using the Zoom tool on the Viewport Navigation Tools toolbar.

14. In the Align Selection dialog, place a checkmark in the Z Position checkbox.

15. Set the Current Object to Maximum, and the Target Object to Minimum.

16. Click the Apply button.

The spotlight should move up the height of the cone geometry.

17. Remaining in the Align Selection dialog, place a checkmark in the both the X and Y Position checkboxes.

18. Set both the Current and Target objects to Center and then click the Apply button.

19. Click OK to complete the Alignment.

Array

The Array tool is a bit complex, but very powerful. It will allow you to transform a selection of objects in one, two or three axes simultaneously. Let's do a simple array to make two copies of the light that we added.

20. Make sure that the light is selected, and then choose **Array** from the Tools menu.

In the Incremental area, you set the spacing, rotation, and/or scale of each transformation. For instance, we will use a 2′-0″ increment in the Y direction. This will create the next arrayed object, 2 feet from the first. If you instead wanted to input the total distance to array along, you would click the small arrow to shift over to the Totals fields. Then you would type in a total value and your arrayed items would fill this total in equal increments.

21. In the Incremental Y column, input **2′-0″**.

For the Type of Object, we will use the default of Instance (more on this below). For the Array Dimensions, this will be a 1D array—the objects are copied along a single axis.

22. In the 1D field, input **3** for the Count and then click OK (see Figure 2–36).

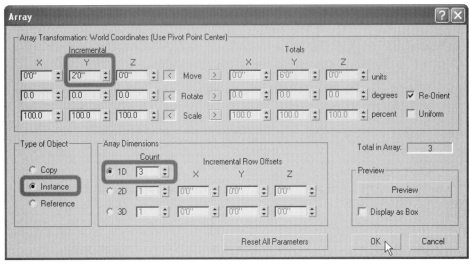

Figure 2–36 *Create a 1D Array*

Try performing another array. Select the bowl this time. Set the incremental Y distance to –3'-0", and the Incremental X and Y scales to **150**. Set the 1D Count to **3**, and the 2D Count to **2**. Set the X Incremental Row Offset next to 2D to **4'-0"**. Click OK to see the result (see Figure 2–37).

Figure 2–37 *Create a 2D array that scales as it copies*

Mirror

The mirror tool allows you to flip objects along an axis and optionally transform them as they are flipped.

23. Using the CTRL key, select all six bowls.

24. From the Tools menu, choose **Mirror**.

    Experiment with the six radio buttons in the Mirror Axis area.

25. Choose the Y-axis radio button.

26. Click and hold down the mouse over the double arrow icon next to the Offset field.

    This is referred to as a "spinner."

WORKING WITH SPINNERS

Spinners are the up and down arrows that allow adjustment to the values for nearly all settings in VIZ Render. If you click and hold down the spinner, it will adjust at an average rate. If you hold down the CTRL key while dragging a spinner, it will change the value at a faster rate. Holding the ALT key, while dragging a spinner will change the value at a slower rate. Right-clicking while adjusting the spinner will reset the value to what it was set at before you began adjusting the value. Right-clicking at any other time will reset the spinner to the lowest allowable value.

27. Drag the spinner up and down to interactively see the effect of the Mirror offset. (see Figure 2–38).

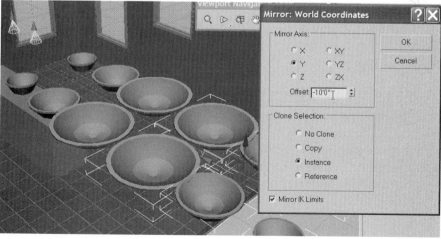

Figure 2–38 *Mirror a copy of the selection along the Y-axis*

Like the Array, the Mirror command has various Clone options. For now choose Instance and then click OK.

CLONE

Cloning is copying. There are three types of clone operations: Copy, Instance and Reference. A Copy is a straight copy. Both the original and new object are completely separate and unique objects. When an Instance is created, it and the original remain linked together. Changes made to any of the instances occur on all instances. A Reference is similar to this. However, only changes to the original object are passed down to the references. Changes made to one of the references are not passed to the original or any other references. Another way to think of it is that instances are all equal to one another. A change to one, changes all. With a reference, there is a single "parent" object whose changes are applied to the references, but not the other way.

In addition to the Array and Mirror commands, which both had cloning options, you can clone objects using the basic transforms (Move, Rotate and Scale) and the SHIFT key.

1. Undo the mirrored and arrayed bowls, leaving only the original.

 TIP If you have performed other operations and Undo is not practical, select all of the copied bowls with the CTRL key (don't select the original) and then press the DELETE key.

2. Select the one remaining bowl, right-click, and choose **Move** from the Edit/Transform Quad menu.

3. Hold down the SHIFT key, and click and drag the Y-axis constraint of the Move Gizmo.

   Release the mouse once you have moved it away from the original object. The Clone Options dialog will appear.

4. In the Object area, choose **Instance**.

Notice that you can optionally change the number of copies to be greater than one. If you do this, the distance that you dragged when you released the mouse will be used as the incremental spacing between the cloned objects. This gives you functionality similar to the 1D Array covered above. Another option in this dialog that you do not have with either Array or Mirror is the ability to indicate the name of the cloned objects.

5. Accept both defaults for Number of Copies and Name and then click OK.

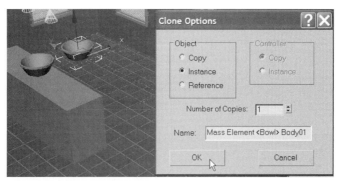

Figure 2–39 *Create a single Instance of the bowl using the* SHIFT *key clone method*

The bowl thus created is an instance. However, because of the limited modeling abilities in VIZ Render, it is difficult to see the impact of your choice of clone type. However, earlier, we cloned the spot lights with the array command. Let's make a simple adjustment to one of them, and because they are also instanced, it will apply to all three.

6. Select one of the spotlights, and on the Modify Panel, expand the Spotlight Parameters rollout.

7. Begin dragging the Falloff/Field spinner. (Do not release the mouse yet.)

Notice that the cone representing all three spotlights is dynamically adjusting to the spinner.

8. Right-click the mouse to cancel the edit without applying it.

Select the bowl again, right-click and switch to rotate, and using the SHIFT key again, clone one or more copies. This time, each copy will incrementally rotate rather than move. You can try the same thing with scale if you wish. Once you have finished experimenting with the various cloning options, delete all but the original bowl.

9. Save the scene.

SELECTION FLOATER

The Selection Floater is a dialog that lists all of the objects within a scene by name. You can select items from the list and select them in the scene. This is very useful in large complex scenes with lots of objects.

1. From the Tools menu, choose Selection Floater, (or click the Selection Floater icon on the Selection toolbar).

Scroll through the list of objects.

ADT Objects from your merged scene will have names like: **Wall <Standard>**. In addition, subcomponents of ADT objects will also be visible. So indented beneath

Wall <Standard> might be something like **Wall <Standard> Unnamed**. It is possible to rename objects in your scene if you prefer. Do this on the Modify panel with an object selected. To select an object using the Selection Floater, click its name in the list (you can use the SHIFT and CTRL keys to select multiple objects), and then click the Select button in the bottom right of the dialog (see Figure 2–40).

2. Make a selection using the Selection Floater.

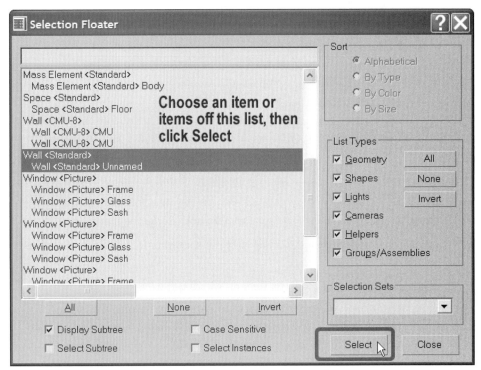

Figure 2–40 *Using the Selection Floater to select an object or objects*

Notice that the object(s) are selected in the scene and that the Selection Floater remains open. The Selection Floater is modeless and you can keep it open onscreen as you work if you wish. Beneath the list of object names are three buttons: All, None, and Invert.

3. Keeping whatever you have selected in the list highlighted, click the Invert button.

This will reverse the selection to highlight everything accept what was previously highlighted. Click Select if you wish to apply the selection to the scene.

4. Click the All button, and then click the None button.

These two buttons do exactly as you would expect.

5. Remove the Display Subtree checkmark.

This will remove all indenting and hierarchy from the list. However, this makes the list hard to read. It is recommended that you leave this enabled.

6. Return the checkmark to the Display Subtree box and then place a checkmark in the Select Subtree box.

7. Click on one of the Window <Picture> objects.

Notice how all of the nested subcomponents of the highlighted window also highlight. This tool is useful if you want to edit the entire ADT object and not just one of its components.

8. Place a checkmark in the Select Instances checkbox and then click on one of the Light objects at the top of the list.

Their names are: FPoint01, FPoint02, and FPoint03.

Notice that highlighting one of them will highlight them all. When you clone Instances (see above) you can use this option to select them all.

On the right of the dialog is a series of checkboxes that enable you to filter the list of selectable objects by type. Geometry would include all of the objects linked over from ADT. Cameras, Lights and Helpers are objects created in VIZr. Shapes and Groups/Assemblies are special classes of objects that sometimes occur in a VIZr scene. For instance, the Daylight System that we created in Chapter 1 would be considered a Group/Assembly. To use these filters, remove the checkmark from the items that you want to filter out. Use the None, All and Invert buttons in the same way as we did for the object name list above.

Finally, there is a text field at the top of the Selection Floater. You can use it to type in partial names for the items that you wish to select using standard Windows wildcards. For example, perhaps you wanted to select all of the Wall objects. You can type "*Wall*" in the filter field and all of the objects that contain the word "Wall" in their name will highlight. Use the "Case Sensitive" checkbox if you wish the filter field to treat "w" as different than "W."

9. In the filter text field at the top, type: ***wall**.

Notice that only the Wall objects highlight.

10. In the filter text field at the top, type: ***glass** (see Figure 2–41).

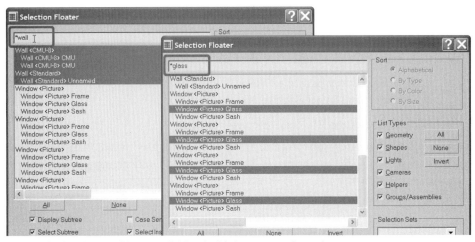

Figure 2–41 *Use the Filter text field to highlight items with similar names*

11. Close the Selection Floater and then Save the scene.

RENDERING AND RADIOSITY

We will only touch upon the high level concepts of rendering and radiosity here in this section. For detailed explanations and techniques, refer to the projects in Chapters 5 through 9. For this sequence, our primary goal is to convey the most important concepts behind rendering and radiosity.

Merge the Remaining Walls and Ceiling

Before we render the scene, we should enclose our stage set here and make it more of a room. We will merge in an existing DRF file that was created for this purpose. Merge was covered above.

1. From the File menu, choose **Merge**.

2. Browse to the *C:\MasterVIZr\Chapter02* folder, select the file named *Finish Room.drf* and click Open.

3. Beneath the object list in the Merge dialog, click the All button to select all objects and then click OK.

Two new walls and a ceiling will appear enclosing our room. The merged file also contains a camera. We can set the viewport to look through this camera so that we can see into the room again.

4. Make sure the Perspective viewport is active and then press the "**C**" key on the keyboard.

5. From the Rendering menu, choose **Render** and then at the bottom of the dialog, click the Render button. (see Render Stats 02.01).

Render Stats 02.01

Render Time: 13 seconds
Radiosity Processing NA
Music Selection: "Love Me Do" The Beatles – The Beatles 1 (2000)
Color Plate: NA
JPG File: Rendering 02.01.jpg

Not a very exciting rendering. We have lots of light on the bowl and countertop and everything else is pitch-black. However, even if these three spotlights were the only lights in the scene, the surfaces would reflect some light back into the scene and indirectly light the neighboring surfaces. This natural bouncing of light and indirect light is simulated in VIZ Render by a process called Radiosity.

Radiosity

Radiosity solutions will be generated for each of the renderings in this book. Any believable rendering you see today makes use of one of two methods of lighting: "Fakeosity" or Radiosity. Radiosity is the process of calculating bounced light throughout a scene. *Fakeosity* is the manual process by which supplemental lights are placed throughout the scene to light dark spots such as under a table, in a dark corner, or within covered spaces in an exterior scene—faking the behavior of real lighting.

6. Click the Clone Rendered Frame Window icon at the top left of the render window.

   This makes a copy of the window so we can compare the changes we make for the next rendering.

7. From the Rendering menu, choose **Render** again.

8. At the bottom of the dialog choose **Draft with Radiosity** from the Preset list.

9. In the Select Preset Categories dialog, click the Load button.

10. In the Render Scene dialog, click the Render button (see Render Stats 02.02).

Render Stats 02.02

Render Time: 15 seconds
Radiosity Processing Time: 4 seconds
Music Selection: "Butterfly" Lenny Kravitz – Mama Said (1991)
Color Plate: NA
JPG File: Rendering 02.02.jpg

Not exactly a competition winner but definitely an improvement. In the first image the only light we saw was that which hit a surface directly from the light source. With Radiosity, we calculated the light being bounced all about the room; off the counter to the ceiling and back again to either the walls or the floor. The scene is still quite dark, but we are now starting to see some detail in the walls, windows, and ceiling.

With a few more changes, this rendering may not look half bad even for a chipboard model.

11. In the Render Scene dialog, open the Preset list and choose: **Production with Radiosity**.

 NOTE *This Preset and the one used in the previous rendering are provided along with two others—Draft without Radiosity and Production without Radiosity—out-of-the-box. The Production presets will typically achieve better quality.*

12. In the Select Preset Categories dialog, be sure that everything is selected and then click the Load button.

If an Advanced lighting plug-in warning dialog appears, simply click OK.

As we noted above, the Radiosity solution is doing a nicer job on the lighting, but the overall effect is not being seen throughout the rendering because it is too dark overall. Let's adjust two settings to produce a better result.

13. In the Render Scene dialog, click the Radiosity tab and then expand the Radiosity Meshing Parameters rollout.

14. In the Global Subdivision Settings area, place a checkmark in the Enabled checkbox and then set the Meshing size to **1'-0"**.

Radiosity applies a mesh to the surface of all objects. When the Global Subdivision setting is enabled, the Meshing size parameter controls the smallest size of the mesh components. Radiosity stores bounced light information at each of the vertices on the mesh, therefore, the finer the mesh (smaller size) the higher the quality the solution will be. However, very small mesh sizes will also have the potential to increase processing times and file sizes dramatically.

15. Click on the Environment tab and in the Logarithmic Exposure Control Parameters rollout, increase the Brightness to **80** (see Figure 2–42).

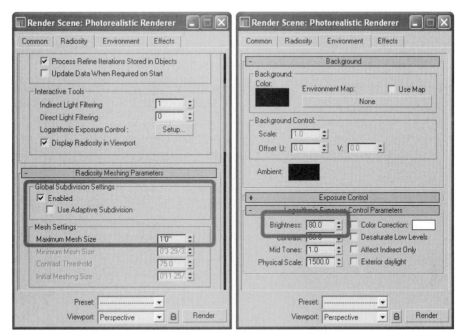

Figure 2–42 *Change the Global Subdivision Settings and the Brightness settings*

16. In the Render Scene dialog, click the Render button (see Render Stats 02.03).

 Render Stats 02.03

Render Time: 2 Minute 12 seconds
Radiosity Processing Time: 53 seconds
Music Selection: "Don't Know Why" Norah Jones – Come Away With Me (2002)
Color Plate: NA
JPG File: Rendering 02.03.jpg

It is still not an award winner, but in a slightly longer period of time the results are much better. We now have some tonal variation on the ceiling and can see the color of the walls. The lighting in the bowl is a bit washed out, but not necessarily unrealistic. Feel free to experiment further with this dataset if you wish. We will get into much greater detail on Radiosity and rendering in the chapters that follow, starting in Chapter 5. In particular, Chapter 6 covers Radiosity settings in much greater detail.

View the Radiosity Mesh

Let's take a look at what the Global Subdivision Settings did to the model. To view the Radiosity meshing, view the model in wireframe.

17. Right-click the Camera:Camera01 viewport label and choose **Wireframe**.

18. Minimize the current viewport and then repeat in any other viewports that are not already displaying wireframe.

19. Zoom Extents all on Viewports (see Figure 2–43).

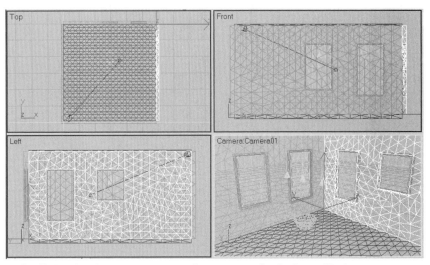

Figure 2–43 *Display the viewports in wireframe to see the Radiosity Meshing*

 TIP If you are concerned about large file size, it's generally a good idea to reset the Radiosity solution before saving. To do this, open the Render Scene dialog and click the Rest All button on the Radiosity tab.

20. On the Radiosity tab of the Render Scene dialog, click the Reset All button to reset the Radiosity solution.

21. Close VIZ Render and Save the scene.

I-DROP

Autodesk's i-drop components are invaluable in rendering. From Autodesk's own website: "Autodesk i-drop® technology gives you the ability to drag-and-drop content from the web straight into your drawing session. It also gives designers and developers the power to create web pages that can easily be dragged-and-dropped into i-drop–aware Autodesk design products. One component of this technology is the i-drop Indicator, that modifies your browser behavior so that you can initiate the drag-and-drop action."

 NOTE Please see Appendix B for links to Autodesk's i-drop web site and other manufacturer sites that offer i-drop content.

We will use i-drop components from various manufacturers' web sites throughout this book. When available, i-drop content is much easier to access and incorporate into your scenes. It has the immediate benefit of accurately representing the real items that you intend to specify in your designs. The process to use them could not be simpler.

i-drop from the Herman Miller Web Site

1. Launch VIZ Render.

   Do not open an existing scene. We will work in the empty scene loaded by default.

2. Using your web browser, browse to: *http://www.hermanmiller.com*.

3. Click the link to: *Architects & Designers*, then: *Planning and Visualization* and finally: *3D Models*.

TIP The URL to this web address (and all web addresses referenced in this book) is included on the CD ROM in the *C:\MasterVIZr\Appendix B* folder.

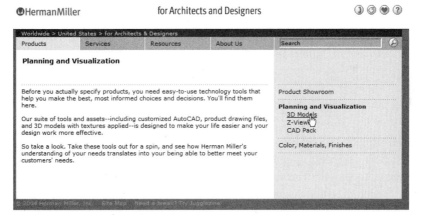

Figure 2–44 *Herman Miller's 3D models location on their web site*

Herman Miller web site and i-drop content used with permission from Herman Miller, Inc.

   Move your mouse over the image labeled Aeron Work Chair. If you do not see the i-drop icon, follow the directions for installing the i-drop client.

4. From the dropdown list on the right, choose **Lounge Seating**.

5. Scroll down to **Goetz Sofa**.

6. Move your cursor over the image (the eyedropper icon should appear), and click and drag to drop it into your VIZr scene (see Figure 2–45).

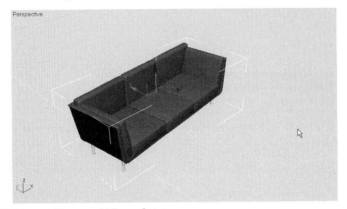

Figure 2–45 *i-drop a Herman Miller sofa into your scene*

7. From the context menu that appears in VIZr, choose **Merge File**.

 NOTE If you get an "Obsolete Data" dialog, simply click OK.

The couch may not be located directly on your cursor.

8. Zoom Extents, and then from the File menu, choose **Save**.

9. Browse to the *C:\MasterVIZr\Props* folder.

10. In the File name field, type: *MVIZr-CH02 - Living Room Couch* and then click Save.

We have also provided a local web site that you can use to access the same content if you do not have an internet connection. Using your Content Browser, add a catalog and browse to the *C:\MasterVIZr\Catalog* folder. Change the Files of type to HTM files. Select the *Index.htm* file and then click Open. Browse this catalog to find all of the Herman Miller and Scott Onstott content used in this book.

There are several other items from Herman Miller that we will use in the upcoming chapters. You can create more of them now by following the same process, or simply wait untill you read those chapters. If you wish to experiment further, i-drop the following into new VIZr scenes following the same steps outlined here (see Color Plate C–3b for examples):

▶ Eames Lounge Chair

▶ Eames Walnut A Shape

▶ Eames Walnut B Shape

The complete collection of i-drop items can be seen in the Color Plates section in Color Plate C–3b.

THIRD PARTY PLUG-INS

Several third-party plug-ins are available for Autodesk Architectural Desktop and VIZ Render. In particular, two specifically for VIZ Render will be showcased in the tutorials in this book. The plug-ins themselves are free to download and install. However, beyond the sample content provided, these companies charge a fee to purchase additional content. The two plug-ins in question are: RPC by Archvision, and EasyNat by Bionatics. RPC stands for "Rich Photorealistic Content." These items are placed into your VIZr scenes like other objects. However, the geometry thus added is comprised of very simple planes with low face counts. Mapped to these surfaces are high-quality photorealistic texture maps that automatically adjust to the camera angle from which they are viewed. Bionatics provides similar content for plants and trees. These items are low polygon count trees and plants whose parameters can be adjusted directly in VIZr.

It is not required that you install these plug-ins to complete the tutorials in this book. However, to get the most out of each lesson, you are encouraged to do so. To get the latest information available on each plug-in, open the content browser, (there is an icon for it on the Render toolbar, or you can access it from the Windows Start menu). On the main page, click the Architectural Desktop & VIZ Render Plug-ins catalog. Then follow the links to RPC and Bionatics. There are several other plug-ins listed that may be of interest to you. You are encouraged to check these out at your leisure.

If you do not have a live Internet connection, you can find the installers for these plug-ins with instructions to install them on the Mastering VIZ Render CD ROM in the *Plug-ins* folder. (see Figure 2–46).

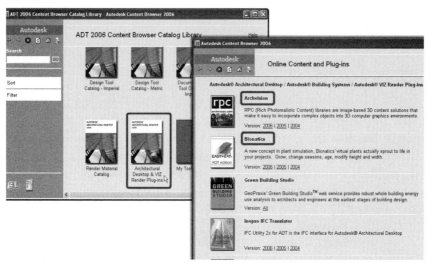

Figure 2–46 *Third-party plugins available for ADT and VIZr in the content browser*

Archvision RPC Plug-in and Bionatics EasyNat Plug-in provided for use here with permission from Archvision and Bionatics respectively.

CONCLUSION AND SUMMARY

In this chapter we have explored a wide variety of tools that are critical to understanding the VIZ Render interface and work flow. Refer back to the sections in this chapter as you work through the upcoming tutorials in the remaining chapters. Among the topics that we covered, are the following:

There are several types of models, both 2D and 3D, created in today's CAD software.

Autodesk Architectural Desktop and Autodesk VIZ Render work together to provide a complete 3D modeling and rendering solution.

Models are created within Architectural Desktop and then linked to VIZ Render to create presentations.

There are three major components the 3D rendering process: Build the Model, Add Lights and Materials, and then Create the Rendering. Each requires different but equally important skills and procedures.

The VIZ Render User Interface share some similarities to ADT, but presents many unique aspects as well.

The File Link from ADT is dynamic and can be reloaded with a variety of options.

Items from other VIZr scenes can be merged into the current scene with ease.

Many useful tools like Align, Array, and Mirror make exact positioning and cloning of merged items simple.

Cloned (copied) objects can be instanced or referenced to remain linked to the original object.

Radiosity provides a great deal of realism to a scene by simulating bounced light off of scene surfaces.

The process of rendering is iterative with each rendering's quality being adjusted by sometimes subtle variations of settings.

Content from manufacturers can be easily imported into a VIZr scene via Autodesk i-drop.

SECTION II

The Projects

This section is devoted to a series of projects with specific goals. In each chapter, you will be introduced to a dataset and given a description of its contents. The goals of the chapter will also be outlined in the introductory passage. By following a particular goal—such as generating an interior rendering using daylight—you will be introduced to a variety of skills and tasks needed to complete that task. Each chapter includes one or more color images in the Color Plates section, and each Rendering file is provided on the CD.

Section II is organized as follows:

CHAPTER 3

Models – Part I

INTRODUCTION

Most of the chapters in this book will explore the various aspects of rendering and presentation from the point of view of several sample project scenarios. Before any rendering can take place, however, we must have a model from which to generate our presentations. In most of the upcoming chapters, the majority of the modeling will be provided in the datasets accompanying those chapters. In this chapter, however, the task will be to build a model. This model will then serve as our dataset in some of the upcoming chapters. The task of building the model is a lengthy one. Therefore the complete process is broken into two chapters. This is Part 1; Part 2 will be presented in Chapter 4.

OBJECTIVES

Although it is assumed that most readers have a basic understanding of modeling techniques in ADT, stepping beyond the basic out-of-the-box content can yield terrific results. We will explore several advanced modeling techniques in Autodesk Architectural Desktop within this chapter and the next chapter. Among the topics covered in this chapter will be the following:

- Import existing Styles and Definitions
- Build and modify complex Wall Styles
- Work with complex Door and Window Styles
- Assign materials to objects within Architectural Desktop
- Work in a Project Navigator project environment

THE TOWNHOUSE DATASET

The project for this chapter is a two-story townhouse. The model is provided as a series of Constructs in the ADT Project Navigator. All of the Constructs have been provided except the façade of the townhouse. In this chapter we will focus on modeling of the townhouse façade. (Please refer to the "Models" heading in Chapter 2 for more discussion on the widely used term "Model.") The actual model pieces created herein will be fairly simple, with some rather sophisticated accoutrement added. Since this is a townhouse, the focus of this rendering project is on the front façade of the building. However, due to the nature of the ADT Project Navigator and Display System structure, we will slice this multistory façade wall into pieces; one for each floor of the project. This follows standard ADT procedures, where each floor of an ADT project is modeled in a separate file and then "stacked" up in Project Navigator to form a complete composite model. The basic structure of this model includes the following features, which will be addressed in each of the headings to follow:

Chapter 3 will cover:

- ▶ Complex Wall Styles containing horizontal and vertical components used as the primary façade Walls.
- ▶ Custom Door & Window Styles: custom settings and Muntin Display.

Chapter 4 will cover:

- ▶ Custom Door & Window Styles: custom Endcaps and Mass Element Custom Block Display.
- ▶ Custom moldings: simple band components, Wall Sweeps, Body Modifiers, and 3D Endcaps.
- ▶ Custom Structural Member Styles used as moldings.

FRONT FAÇADE WALLS

Let us first establish the basic geometry needed to convey our design. By starting out with a few simple Walls, Doors, and Windows, we can get a very good idea of how the overall design will look in terms of structure, form, and scale. We can then progressively refine each of the components to detail them to a sufficient level. The final result is shown in Figure 3–1 on the left, with a detailed Wall Section showing the makeup of the Wall Styles that we will build on the right. You can refer to Figure 4–55 in Chapter 4 and Color Plate C–4a to see the final result as well.

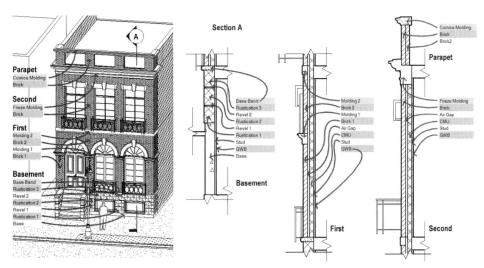

Figure 3–1 *Townhouse completed front façade (shown as it will appear after Chapter 4)*

Start Autodesk Architectural Desktop and Load the Townhouse Project

1. If you have not already done so, install the dataset files located on the Mastering VIZ Render CD ROM.

   Refer to "Files Included on the CD ROM" in the Preface for information on installing the sample files included on the CD.

2. Launch Autodesk Architectural Desktop from the icon on your desktop.

   You can also launch ADT by choosing the appropriate item from your Windows Start menu in the Autodesk group. In Windows XP, look under "All Programs"; in Windows 2000, look under "Programs." Be sure to load Architectural Desktop, and not VIZ Render, for this exercise.

3. From the File menu, choose **Project Browser**.

4. Click to open the folder list and choose *My Computer* and then your C: drive.

5. Browse to the *C:\MasterVIZr* folder.

6. Double-click **Townhouse03** to make this project current (you can also right-click on it and choose **Set Current Project**). Then click Close in the Project Browser.

 IMPORTANT If a message appears asking you to re-path the project, click Yes. Refer to the "Re-Pathing Projects" heading in the Preface for more information.

Create Basement Facade Walls

Several Constructs have already been created for this project. There is one for each floor including the parapet at the roof, a couple for the surrounding site information, and other miscellaneous items. Constructs are the building blocks of ADT projects. Each Construct represents a unique portion of a building model. Like a jigsaw puzzle, when all the Constructs for a project are assembled together, you have a complete building model. For more information about ADT Drawing Management, Constructs, Elements, Views, and Sheets, see the online help, or pick up a copy of *Mastering Autodesk Architectural Desktop* by Paul F. Aubin. These topics are covered in more detail in those references. Let's begin with the Basement Construct and work our way up.

> When you loaded the Townhouse project with Project Browser above, the Project Navigator palette should have opened automatically. If it did not, you can load it from the Window menu.

7. On the Project Navigator Palette, click the Constructs tab and then double-click to open the *Basement* Construct to open it.

The file contains two walls and a slab (the dashed blue lines). To add the façade wall in the correct spot, we will draw it relative to this existing geometry. Perhaps one of the most important decisions early in a project is the Baseline (or reference point) used for the Walls. Naturally, you can always change this point later if necessary, but it is always better if you can pick a good location for the Wall's Baseline from the very start. In this project, the Basement Walls are made from CMU, while the upper Walls are CMU and Brick veneer. The default location for the Wall Style Baseline is typically the point that separates structural and nonstructural components, such as the face of CMU in the air gap between the CMU and Brick. If you select one of the existing Walls in this file, you will note that the Baseline has been configured to be near the center of the Wall. It is actually 5 1/2″ from the inside face of the Wall. This Baseline reference point will keep the Baseline in the same position for all floors, even though the thickness of CMU varies from one floor to the next.

Using the existing geometry as a guide, we will use the Baseline Justification option and the Baseline Offset feature to position the Walls exactly where required.

8. Select either of the existing Walls, right-click, and then choose **Add Selected**.

9. On the Properties Palette, verify that the Justify is set to **Baseline**. In the Baseline Offset field, type **5 1/2″**.

10. Begin on the right side, and snap to the point on the blue slab line where it meets the existing Wall.

11. Move down and snap to the corner of the Slab.

12. Continue to the left and snap to the next corner of the slab, and then finish by snapping to the point where the Slab meets the Wall on the left.

13. Press ENTER to end the command (see Figure 3–2).

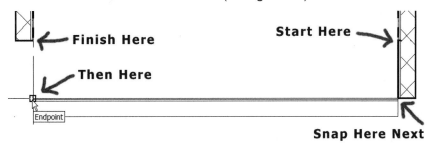

Figure 3–2 *Draw three Wall Segments snapping to the Slab as you Draw*

14. Select all three of the Walls just drawn, right-click, and choose **Copy Wall Style and Assign**.

15. On the General tab, change the name of the Style to **Townhouse Basement Facade Walls**, and then click **OK** to close the Wall Style Properties dialog.

16. **Save** the file.

Create First Floor Facade Walls

Following a similar procedure, we will draw some Walls in the *First* Floor Construct.

17. On the Project Navigator Palette, click the Constructs tab and then double-click the *First Floor* Construct to open it.

The file also contains two exterior alley walls and a slab (the dashed blue lines). It also contains a few other walls and some stairs. Rather than draw the façade walls in this file with a Style that is already resident in the file, we will use one from the Content Browser library instead.

18. On the Window menu, choose **Content Browser**.

TIP You can also click the Content Browser icon on the Navigation toolbar or press CTRL + 4 within ADT, or find it on the Start menu in the same group as your version of ADT (see Figure 3–3).

Figure 3–3 *Launch the Content Browser using the toolbar icon*

As its name implies, use the Content Browser to *browse* content contained on your system. Several catalogs appear in the default installation of ADT. You will find both AEC Content (Blocks, Multi-View Blocks, and Custom Routines) and ADT Object Styles among the content provided.

19. Locate the catalog named **Design Tool Catalog – Imperial** and click on it (see Figure 3–4).

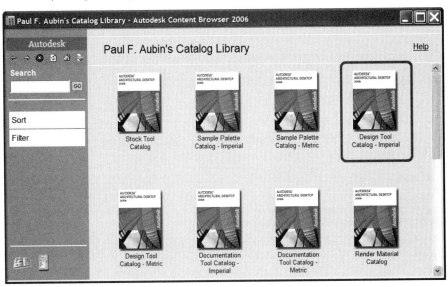

Figure 3–4 *Open the Content Browser to locate a specific Wall Style*

20. Using the scrolling panel at the left, locate the **Walls** category and click on it.

   This should reveal five Wall Style categories organized by construction type.

21. Click on the **CMU** category to browse it.

There are several pages of CMU Wall Styles available to browse. If you have left the default display settings in place for your Content Browser, you will see five rows of Style icons per page and a total of five pages of CMU Wall Styles.

 TIP You can edit the number of rows displayed on the home page of the Content Browser by choosing the Library View Options icon in the lower-left corner of the Content Browser (see Figure 3–4).

22. Click **Next** as required to locate the Style named **CMU-4 Air-2 Brick-4 Furring**. (In the default view settings, this Style is located on page 2.)

23. Click and hold the left mouse button down over the *i-drop* (eye-dropper) icon and wait for it to "fill up."

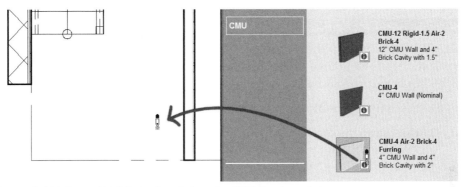

Figure 3–5 *i-drop a Wall Style directly into the ADT drawing window*

24. Keeping the mouse button depressed, drag the eye-dropper into the ADT drawing window. Release the mouse button within the ADT window to execute the Add Wall command with the CMU-4 Air-2 Brick-4 Furring Style active (see Figure 3–5).

25. On the Properties palette, change the Base Height to **12′-6″** and the Justify to **Baseline**.

26. Set the Baseline Offset to **5 1/2″** and the Roof line offset from base height to **1′-0″**.

27. Draw three Walls in the same manner used above for the Basement Walls (see Figure 3–6).

28. Press ENTER when finished to end the command.

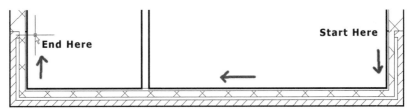

Figure 3–6 *Draw three Walls relative to the Slab edge as done above for the Basement Walls*

29. Select all three of the Walls just drawn, right-click, and choose **Copy Wall Style and Assign**.

30. On the General tab, change the name of the Style to **Townhouse First Floor Facade Walls**, and then click **OK** to close the Wall Style Properties dialog.

31. **Save** the file.

Create Second Floor Facade Walls

Again, we will move on to the next floor before editing the façade wall of the first floor any further. The Second Floor is very similar to the First. Therefore, we can copy the Walls that we have added here up to the second floor.

32. In the *First* Floor Construct, select the three façade Walls.

33. Hold down the CTRL key and drag these Walls to the Project Navigator Palette and release them on the *Second Floor* Construct (see Figure 3–7).

 TIP Be sure to drag from the highlighted edge of the Walls; do not select any of the Grips while dragging.

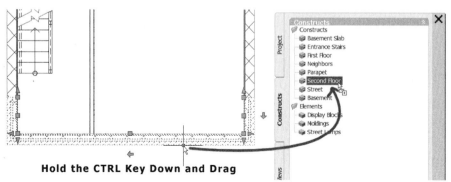

Hold the CTRL Key Down and Drag

Figure 3–7 *Drag and Copy the façade Walls from the First Floor to the Second Floor*

Using the CTRL key while dragging makes a copy of the dragged geometry in the other Construct. If you wish to move something to the other Construct, don't use the CTRL key.

34. On the Project Navigator Palette, click the Constructs tab, and then double-click the *Second Floor* Construct to open it.

Notice that the three Walls that we just dragged are here in the Second Floor file in the correct physical location.

35. Select one of the three façade Walls in the Second Floor file, right-click, and choose **Edit Wall Style**.

36. On the General tab, change the name of the Style to **Townhouse Second Floor Facade Walls**, and then click **OK** to close the Wall Style Properties dialog.

37. **Save** the file.

Create Parapet Fa,ade Walls

Finally, we will repeat the process used above to copy the Second Floor Walls to the Parapet Construct. We will then make a few adjustments to the Walls as required by the Parapet file.

38. Using the CTRL key drag technique, repeat the same process used above to copy the three façade Walls from the *Second Floor* Construct to the Parapet Construct.

39. On the Project Navigator Palette, open the **Parapet** Construct.

40. Select all three of the Walls just drawn on the Properties Palette, and change the Base Height to **3'-10"**.

41. With the same three Walls still selected right-click and choose **Copy Wall Style and Assign**.

42. On the General tab, change the name of the Style to **Townhouse Parapet Facade Walls**, and then click **OK** to close the Wall Style Properties dialog.

43. **Save** the file.

FRONT FAÇADE FENESTRATION

Now that we have our Walls in place on each floor, let's add some Doors and Windows to the façade. We will add a fenestration to each floor: Windows at the Basement, Doors at the first floor, Windows at the second floor, and some openings in the parapet at the roof level.

We will use very generic Door and Window Styles for now, this way we can assemble all of the pieces and have a look at the overall façade before embellishing it further.

Create Basement Windows

We will begin with the Basement as we did above. The Basement needs two Windows.

1. On the Project Navigator Palette, click the Constructs tab and then double-click to open the *Basement* Construct to open it.

NOTE If you left the *Basement* Construct open above, then this action will simply make that file active.

2. On the Design tool palette, click the **Window** tool.

3. On the properties palette, change the Width to **4'-8"** and the Height to **2' 10"**. Make sure that "Measure to" is set to **Outside of Frame**.

4. Scroll down on the properties palette and find the Location grouping. Change the "Position along wall" to **Offset/Center**.

5. Change the "Vertical alignment" to **Sill** and set the "Sill height" to **8"**.

6. At the "Select wall, space boundary" prompt, click the horizontal façade Wall.

   Move the mouse around a bit and notice that with the "Offset/Center" setting the Window snaps easily to the midpoint of the Wall.

7. Click to place a Window in the middle of the Wall at the bottom.

Notice the dynamic dimensions onscreen as you move your mouse. They show the size of the Window and its current position relative to the nearest Wall endpoints and intersections. You can cycle through these dimensions with the TAB key and then input a value to place the Window with precision. Let's try that now.

   Move the mouse to the right side of the screen so that the next Window we place will be to the right of the one in the middle.

8. Press the TAB key until the dimension between the Window just placed and the new one is highlighted in magenta. With this dimension active, type **2'-"** and then press ENTER (see Figure 3–8).

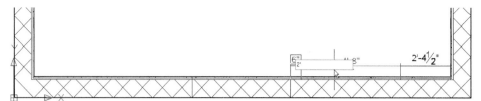

Figure 3–8 *Tab in the dimension and complete the Add Window command after two Basement Windows are placed*

9. Press ENTER again to end the command.

We did not place a Window on the left side of the façade because this is where the steps leading up to the front door will be. As you can see, this gives us two basement windows. However, the Windows do not display currently in Plan view. This is because the Windows fall below the drawing's Display Configuration Cut Plane.

We can edit the Display Configuration Cut Plane, or attach a Cut Plane override to this Wall.

10. Select the main horizontal façade Wall, right-click, and choose **Edit Wall Style**.

11. Click the Display Properties tab and place a checkmark in the Style Override checkbox next to Plan.

12. In the Display Properties dialog that appears, click the Cut Plane tab.

13. Place a checkmark in the **Override Display Configuration Cut Plane** checkbox, and then change the Cut Plane Height to **2′-6″**.

14. Click **OK** twice to return to the drawing (see Figure 3–9).

Figure 3–9 *Change the Cut Plane Height of the Townhouse Basement Facade Walls Style*

15. **Save** the file.

Create First Floor Doors

Next we will add doors to the first floor level. There will be a main entry door on the left, and two French doors directly above the basement windows. By default, Doors have their "Measure to" set to "Inside of frame." This usually is a good choice for generating Door Schedules. However, for this example and for consistency with the Windows on this façade, we will change this default here to "Outside of frame" to match the Windows. First, let's access another style from the Content Browser.

16. On the Project Navigator Palette, click the Constructs tab and then double-click to open the *First Floor* Construct to open it.

 NOTE If you left the *First Floor* Construct open above, then this action will simply make that file active.

17. On the Windows Task Bar, click the Content Browser icon to make it the active application.

TIP You can also use the ALT + TAB key combination to bring it forward.

18. Across the top, just beneath the title of the catalog, click the **Catalog Top** link.

NOTE If you closed the Content Browser, you will need to re-open it (in the same way as outlined in the "Create First Floor Façade Walls" heading above) and browse to the *Design Tool Catalog – Imperial* catalog again.

19. Navigate to **Doors and Windows > Doors**. Locate the Style named: **Hinged – Double – Panel** (on page 3 in the default Content Browser layout) and i-drop it into the ADT drawing window.

20. At the "Select wall, space boundary" prompt, click the main horizontal façade Wall.

21. On the Properties palette, change the Width to **4'-8"** and the Height to **8'-0"**.

22. Set the "Measure to" option to **Outside of Frame**.

23. Move the mouse to the left side of the Wall this time, use the TAB key again, and input a dimension of **1'-0 1/2"** from the left corner of the Wall and then press ENTER.

NOTE This book is written in Imperial units. When using Imperial units, the Architectural unit format is active in ADT where the base unit is inches. Both feet and inches can be entered when you work in Architectural (or Engineering formats) as long as you distinguish feet from inches at the command line and in text input boxes with an apostrophe (') following the number. To enter inches, typing only the number is required. Hyphens are not required to separate feet from inches. Hyphens are used to separate fractions from whole numbers. For more complete information on valid input of Imperial units in the ADT, refer Table 3–1 and the online help.

Table 3–1 *Acceptable Imperial Unit Input Formats*

Value Required	Type This	Or This
Four feet	4'	48
Four inches	4	4"
Four feet four inches	4'4	52
Four feet four and one half inches	4'4-1/2	4'4.5 or 52.5

NOTE Typing the inch (″) mark is acceptable as well: however, it is not required. Dimensions throughout this text are presented in Feet and Inch format for clarity. However, feel free to enter dimension values in whatever formats you prefer. Eliminating the inch mark reduces keystrokes and is recommended despite the inclusion of it in this text. Likewise leading zero as seen in step 23 above can be safely omitted, even though it is shown in this text. Please note also that the default unit format in VIZ Render is not necessarily the same as the linked ADT model.

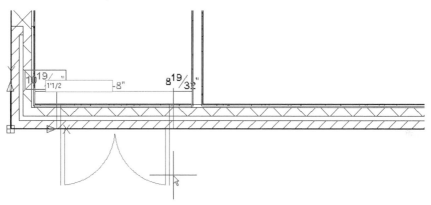

Figure 3–10 *Insert the Hinged – Double – Panel Door relative to the left corner of the façade*

24. On the properties palette, change the Style of the Door to **Standard**.

25. Move the mouse all the way to the right side of the Wall this time, use the TAB key again, and input a dimension of **1′-0 1/2″** from the right corner of the Wall and then press ENTER (see Figure 3–10).

26. Move back toward the center of the Wall (the dimension between this Door and the right corner should be highlighted), input **7′-8-1/2″**, and then press ENTER (see Figure 3–10).

If you place an opening in the wrong place accidentally, do not cancel or press the Undo icon on the toolbar. Instead, right-click in the drawing and choose Undo from the context menu. This Undo will reverse the last Door or Window you placed while the Undo on the toolbar will reverse the entire Door/Window placement command.

The locations and spacing of the three Doors should match the dimensions shown in Figure 3–11. Please note your file does not contain the dimensions, they have been provided here for clarity. If the positions of your Doors don't match the dimensions shown in Figure 3–11, use the Move command or the Location Grip to make the necessary adjustments.

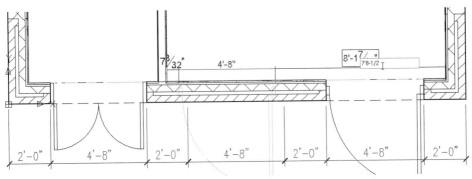

Figure 3–11 *Insert three Doors across the first floor level (dimensions shown for clarity)*

27. Press ENTER again to end the command and then **Save** the file.

Create Second Floor Windows

Next we will add fenestration to the second floor level. On this level, all three openings will be in the same basic locations as those on the first floor, but they will all be the same type. We will use Door/Window Assemblies for these.

28. On the Project Navigator Palette, click the Constructs tab and then double-click to open the *Second Floor* Construct to open it.

 NOTE If you left the *Second Floor* Construct open above, then this action will simply make that file active.

29. On the Design tool palette, click the **Door/Window Assembly** tool.

The settings on the properties palette for this tool will be nearly the same as the Doors and Windows that we added to the first floor.

30. At the "Select wall, space boundary" prompt, click the main horizontal façade Wall.

31. On the properties palette, change the Length to **4'-8"** and the Height to **8'-0"**.

 NOTE The "Width" of a Door/Window Assembly is actually referred to as Length.

32. Scroll down on the properties palette and find the Location grouping. Change the "Position along wall" to **Offset/Center**. Change the "Vertical alignment" to **Sill** and set the "Sill height" to **1'-4"** (see Figure 3–12).

33. Using the "Offset/Center" setting, click to place a Window in the middle of the Wall.

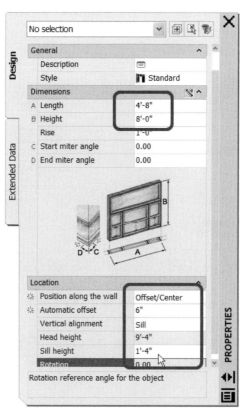

Figure 3–12 *Configure the Properties Settings of the Door/Window Assembly*

34. Use the TAB key function to highlight the dimension between the Assembly just added and the new one, and then type **2'-0″** and press ENTER.

35. Repeat on the other side (see Figure 3–13).

You will end up with three Assemblies. These Assemblies should be lined up directly above the Doors and Windows on the lower levels. The same dimensions used in Figure 3–11 apply to these as well, and can be used to check the placement here.

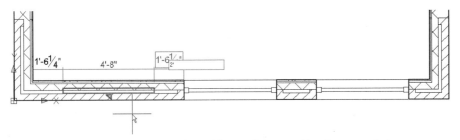

Figure 3–13 *Add the Door/Window Assemblies to the second level using techniques similar to before*

36. **Save** the file.

Create Parapet Openings

With Windows and Doors for the first three levels in place, all we need now is the Openings that penetrate the parapet. Openings are simpler objects than Doors and Windows. They have only dimensional parameters like width and height and do not use styles. However, they are added to the drawing with nearly the identical process as the others.

37. On the Project Navigator Palette, click the Constructs tab and then double-click to open the *Parapet* Construct to open it.

NOTE If you left the *Parapet* Construct open above, then this action will simply make that file active.

38. On the Design tool palette, click the **Opening** tool.

39. At the "Select wall, space boundary" prompt, click the main horizontal Wall again.

40. On the properties palette, change the Width to **4′-8″** and the Height to **2′-8″**.

41. Scroll down on the properties palette and find the Location grouping. Change the "Position along wall" to **Offset/Center**. Change the "Vertical alignment" to **Sill** and set the "Sill height" to **6″**.

42. As before, using the "Offset/Center" setting, click to place an Opening in the middle of the Wall.

43. Use the TAB key function and a **2′-0″** dimension like before to add two more Openings; one on each side of the first. Press ENTER to complete the command.

44. **Save** the file.

As before, these three openings should align with the Windows and Doors below. You can use the dimensions in Figure 3–11 as a guide. The fenestration will get plenty more detailed later on. For now, we have the basic sizes and positions established. Next, we'll move on to refining the design of the Wall Style. But first, let's open up a View file from Project Navigator to see how all these Constructs look when put together. A View file gathers one or more of the building blocks (Constructs) of the project and assembles them at the correct physical locations and heights. In this way, we can easily see a composite of our entire project. A Composite Model View has already been created for this project and is included on the Views tab of the Project Navigator. Be sure that you have saved all floors before continuing.

45. On Project Navigator, click the Views tab, and then double-click the **Composite Model** View file to open it (see Figure 3–14).

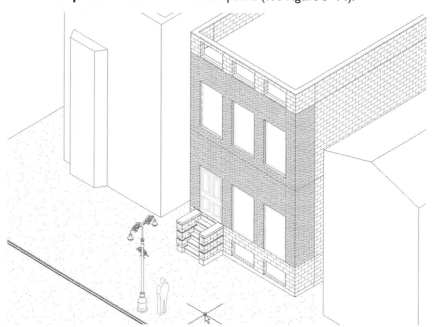

Figure 3–14 *Open the Composite Model View file on the Project Navigator to see how things are shaping up*

BUILDING COMPLEX WALL STYLES

Most often, complex Wall Styles are used to convey detailed construction components primarily in plan. In this example we will have plan components, and we will also use wall components to convey design elements in the façade, such as stone bands and cornices.

Although you can build your styles completely from scratch, the best practice first step to creating a custom Wall Style (or any type of style) is to begin with an existing style that is close to what you wish to create. We have already completed this step by importing the **CMU-4 Air-2 Brick-4 Furring** Wall Style above when building the first and second floors, and by using the existing Walls for the other floors. These styles already include several components and their associated materials. This will greatly reduce the amount of effort we must expend in designing our new styles.

Basement Wall Style Components

Let's get started building our first custom Wall Style. As before, we will begin at the basement level and work our way up. On the base of our façade the design calls for a rusticated base. Therefore, we will edit the Style that was used to draw the façade Walls of the basement to include several stacked block components. As you recall in the "Front Façade Walls" heading above, we already copied and assigned new Wall Styles to the façade Walls in each level. So now, all we need to do is open the files and edit them.

1. On the Project Navigator Palette, click the Constructs tab and then double-click the *Basement* Construct to open it.

 NOTE If you left the *Basement* Construct open above, then this action will simply make that file active.

2. Select the main horizontal facade Wall, right-click, and choose **Edit Wall Style**.

3. Click the Components tab.

As you can see, this Wall Style has four components already. The Wall that we are designing has a rusticated base for the portion that is above grade, and concrete for the portion below grade (see the left Wall Section in Figure 3-1). The Base component is already present in this Style.

Notice that each component has four dimensions that are used to define it: *Edge Offset, Width, Bottom Elevation Offset,* and *Top Elevation Offset.* These four, combined with the Length and Height of the actual Wall object, are used to draw what is essentially a box representing each component of the Wall. Edge Offset is measured from the Baseline of the Wall. It can be either negative or positive, which will offset left or right relative to Baseline. The Baseline of the wall is simply the "zero point" of the Wall. In plan, Baseline varies per Wall Style. In elevation, Baseline is always at Z=0. Width is measured from the Edge Offset and is often positive. However, Width can also be negative and as with Edge Offset, positive and negative simply determines the direction (left or right) in which to measure. Vertical start and end points for components can be measured from one of four points: *Wall Bottom*, *Baseline* (Floor line or Z=0), *Base Height* (Ceiling height), or *Wall Top*. Wall Bottom and Top are the bottom-most and top-most extremes of the Wall, which will obviously vary from one Wall to the next. The exact location of these extremities is controlled by the *Roofline* and *Floorline* options of each Wall object.

The existing CMU component in this style will require some modifications. We will define new components for the other rusticated base pieces. If you look to the left of the Components tab, you will see an interactive viewer embedded into the dialog. As you can see, highlighting a component in the list will highlight that component in green in the viewer. You can also see from the viewer (and component dimensions) that all components currently go all the way from the bottommost point (Wall Bottom) of the model to the top-most (Wall Top). To construct our rusticated base, we must edit the Top and Bottom Elevation values of the CMU.

 NOTE The Wall Style editor is like any other dialog in ADT. You can resize it and adjust the size of each pane in the widow. You should make the Wall Style dialog as large as your monitor will allow and widen the viewer pane as large as you can without causing a horizontal scroll bar to appear. You can also zoom and pan the viewer with your mouse wheel, as well as right-click within the viewer to get a menu of standard viewing options. If you wish to reset the view angle to the default, right-click in the viewer and choose **Preset Views>Left**.

4. In the component list, select the **CMU** component.

5. Rename the CMU component "**Rustication 1**."

6. Change the Top Elevation Offset to **2′-0″** and choose **Baseline** from the "From" list.

When you have finished configuring this component, you will see the component get shorter in the viewer. You will also notice that this component sits directly on top of the Base component and protrudes from it slightly on the left. Next we will duplicate both the Base and Rustication components a few times to complete the rest of the rusticated base of the townhouse façade wall (see Figure 3–15).

 TIP Remember that you can right-click in the viewer to manipulate it, as well as using the wheel on your mouse.

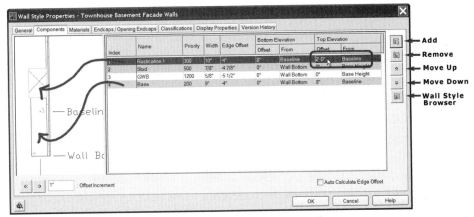

Figure 3–15 *Reconfigure the CMU component as Rustication 1*

7. Select the Base component to highlight it and then click the **Add Component** icon on the right. (It is the one on top; see Figure 3–15.)

This will make a copy of the selected component, in this case, the concrete Base component.

8. Rename the copy **Revel 1** and move it above Rustication 1 (which is labeled *Move up* in Figure 3–15).

9. Change the Bottom Elevation Offset for Revel 1 to **2′-0″** From **Baseline**.

10. Change the Top Elevation Offset for Revel 1 to **2′-1″** From **Baseline**.

You should now have a thin component sitting on top of Rustication 1.

11. Select Rustication 1 and then click the **Add Component** icon.

12. Name the new component **Rustication 2**, and Move it above Revel 1.

13. Set its Bottom Elevation Offset to **2′-1″** and the Top Elevation Offset to **3′ 5″**. (These are both relative to the Baseline, which should already be set.)

14. Repeat this process, copying Revel 1 to create **Revel 2**, and once more copying Rustication 2 to create **Rustication 3**.

15. Move the components as required and set the Bottom and Top Elevation Offsets as shown in Table 3–2.

Table 3–2 *Townhouse Basement Façade Wall Component List*

Name	Priority	Width	Edge Offset	Bottom Offset	Elevation From	Top Offset	Elevation From
Base Band	300	12″	-4″	4′-10″	Baseline	5′-2″	Baseline
Rustication 3	300	11″	-4″	3′-6″	Baseline	4′-10″	Baseline
Revel 2	200	10″	-4″	3′-5″	Baseline	3′-6″	Baseline
Rustication 2	300	11″	-4″	2′-1″	Baseline	3′-5″	Baseline
Revel 1	200	10″	-4″	2′-0″	Baseline	2′-1″	Baseline
Rustication 1	300	11″	-4″	8″	Baseline	2′-0″	Baseline
Stud	500	7/8″	-4 7/8″	0″	Wall Bottom	0″	Base Height
GWB	1200	5/8″	-5 1/2″	0″	Wall Bottom	0″	Base Height
Base	200	10″	-4″	0″	Wall Bottom	8″	Baseline

We need one more component to complete our rusticated base: a stone band to top off the rusticated base (this is shown in Table 3–1, "Base Band"). We will copy one of the rustication components for this.

16. Select Rustication 3 and then click the **Add Component** icon.

17. Rename it **Base Band**, and configure it as shown in Table 3–1. (Note that the Width of this component is **12″**.)

Your Wall Style should now look like Figure 3–16 in the Viewer.

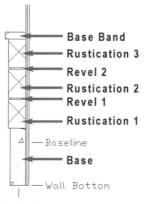

Figure 3–16 *The completed rusticated base in the viewer window (annotations added for clarity)*

18. Click **OK** to close the dialog.

19. From the View menu, choose **3D Views > Front**, and then choose **Shade > Hidden**.

 TIP You can also find these commands on the Views and Shading toolbars, respectively.

If the Grid is displayed, you can turn it off by clicking the GRID toggle button on the Application Status Bar at the bottom of the ADT window. Your file should look like Figure 3–17.

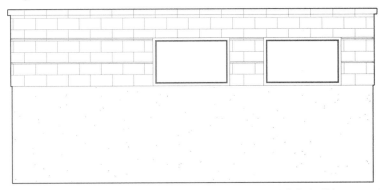

Figure 3–17 *The results of the Townhouse Basement Façade Wall Style Edits viewed from the front*

 NOTE Don't worry about the block coursing yet, we will address this later.

20. **Save** the file.

First Floor Wall Style Components

The First Floor Style is a bit simpler than the Basement Wall. However, we build it the same way.

1. On the Project Navigator Palette, click the Constructs tab, and then double-click the *First Floor* Construct to open it.

 NOTE If you left the *First Floor* Construct open above, then this action will simply make that file active.

2. Select the main horizontal facade Wall, right-click, and choose **Edit Wall Style**.

This Wall Style already has most of what we need. We just need to add two sections of brick separated by a stone band.

3. Select the Stud component and change the Top Elevation Offset to **Base Height**. Repeat for the GWB component.

This will stop these two components at the ceiling line of this floor.

4. Highlight Brick, rename it **Brick 1**, and then change the Top Elevation Offset to **7'-4"** From **Baseline**.

5. With Brick 1 still selected, click the **Add Component** icon.

6. Rename the new component **Brick 2**; change the Bottom Elevation Offset to **8'-2"**, and the Top Elevation Offset to **12'-10"**.

You should now have two stacked brick components with a gap in between. To fill in that gap, we need to define some new components that will have dimensions similar to the brick but different materials. We will be discussing materials in a heading below. For now, let's use one of the brick components as a starting point since at least the Edge Offsets of the stone bands will be the same as those of the bricks.

7. Select **Brick 1** again, click the **Add Component** icon, and name the new component **Molding 1**.

8. Change the Width to **5"**, the Bottom Elevation Offset to **7'-4"**, and the Top Elevation Offset to **8'-2"**.

9. With Molding 1 still selected, click the **Add Component** icon.

10. Rename the new component **Molding 2** and move it above Brick 2 in the component list.

11. Change the Bottom Elevation Offset to **12'-10"** and the Top Elevation Offset to **13'-6"**.

12. Double-check all of your settings against those listed in Table 3–3.

Table 3–3 *Townhouse First Floor Façade Wall Component List*

Name	Priority	Width	Edge Offset	Bottom Offset	Elevation From	Top Offset	Elevation From
GWB	1200	5/8″	-5 1/2″	0″	Wall Bottom	0″	Base Height
Stud	500	7/8″	-4 7/8″	0″	Wall Bottom	0″	Base Height
CMU	300	4″	-4″	0″	Wall Bottom	0″	Wall Top
Air Gap	700	2″	0″	0″	Wall Bottom	0″	Wall Top
Molding 2	810	5″	2″	12′-10″	Baseline	13′-6″	Baseline
Brick 2	810	4″	2″	8′-2″	Baseline	12′-10″	Baseline
Molding 1	810	5″	2″	7′-4″	Baseline	8′-2″	Baseline
Brick 1	810	4″	2″	0″	Wall Bottom	7′-4″	Baseline

NOTE If you were to close the Style Properties dialog at this point, you might notice a strange condition at the ends of the Walls. This is caused by the Endcaps (a set of rules that controls the way a Wall will look at the ends). Don't worry about this now, we will work on the Endcaps later.

13. Click the Endcaps/Opening Endcaps tab.

14. Change the Wall Endcap Style and all Opening Endcaps to **Standard** (see Figure 3–18).

Figure 3–18 *Change all Endcaps to Standard*

15. Click **OK** to close the Wall Style Properties dialog.

16. Change to **Front** view and **Hidden** display as before, and then **Save** the file.

Second Floor Wall Style Components

Let's move to the Second Floor Wall and configure that Wall Style now. We have nearly all the components that we need already. On this floor, we simply need to add a frieze molding.

1. On the Project Navigator Palette, click the Constructs tab and then double-click the *Second Floor* Construct to open it.

 NOTE If you left the *Second Floor* Construct open above, then this action will simply make that file active.

2. Select the main horizontal facade Wall, right-click, and choose **Edit Wall Style**.

3. On the Components tab, select the Stud component, and change the Top Elevation Offset to **Base Height**. Repeat for the GWB component.

4. Select the Air Gap component and change the Top Elevation Offset to 12′ 0″ From **Baseline**. Repeat for the CMU component.

5. Select Brick component, and change the Top Elevation Offset to 12′-0″ From **Baseline**.

6. With the Brick component still selected, click the **Add Component** icon. Rename it **Frieze Molding** (make sure it is above Brick in the component list; move it up if necessary).

7. Change the Width of Frieze Molding to 11″ and the Edge Offset to **-4″**.

8. Set the Bottom Elevation Offset to 12′-0″ and the Top Elevation Offset to 14′-0″.

9. Set the From points of both Top and Bottom Elevation Offsets to **Baseline**.

10. Double-check all of your settings against those listed in Table 3–4.

Table 3–4 *Townhouse Second Floor Façade Wall Component List*

Name	Priority	Width	Edge Offset	Bottom Elevation Offset	Bottom Elevation From	Top Offset	Top From
Frieze Molding	200	11″	−4″	12′-0″	Baseline	14′-0″	Baseline
Brick	810	4″	2″	0″	Wall Bottom	12′-0″	Baseline
Air Gap	700	2″	0″	0″	Wall Bottom	12′-0″	Baseline
CMU	300	4″	−4″	0″	Wall Bottom	12′-0″	Baseline
Stud	500	1 3/8″	−5 3/8″	0″	Wall Bottom	0″	Base Height
GWB	1200	5/8″	−6″	0″	Wall Bottom	0″	Base Height

11. Click the Endcaps/Opening Endcaps tab.

12. Change the Wall Endcap Style and all Opening Endcaps to **Standard** (see Figure 3–18).

13. Click **OK** to close the Wall Style Properties dialog.

14. Change to **Front** view and **Hidden** display as before, and then **Save** the file.

Parapet Wall Style Components

Finally we will configure the Parapet Wall Style. This one is like the second floor, only it needs a molding at the top.

1. On the Project Navigator Palette, click the Constructs tab, and then double-click to open the *Parapet* Construct to open it.

NOTE If you left the *Parapet* Construct open above, then this action will simply make that file active.

2. Select the main horizontal facade Wall, right-click, and choose **Edit Wall Style**.

3. On the Components tab, select the Stud component, and then on the right side, click the **Remove Component** icon (labeled *Remove* in Figure 3–15). Repeat for the GWB component.

4. Using a process similar to above, modify the Wall Style to match the components listed in Table 3–5, including the addition of the Cornice Molding component.

TIP Select the CMU as the starting point for the Cornice Molding, repeating the steps provided for the other floors.

Table 3–5 *Townhouse Parapet Façade Wall Component List*

Name	Priority	Width	Edge Offset	Bottom Elevation		Top Elevation	
				Offset	From	Offset	From
Cornice Molding	300	11″	-4″	3′-2″	Baseline	3′-10″	Baseline
Brick	810	4″	2″	0″	Wall Bottom	3′-2″	Baseline
Air Gap	700	2″	0″	0″	Wall Bottom	3′-2″	Baseline
CMU	300	10″	-4″	0″	Wall Bottom	3′-2″	Baseline

5. Change the Endcaps to Standard as done above, and then click **OK**.

6. Change to **Front** view and **Hidden** display as before, and then **Save** the file.

7. On the Project Navigator Palette, click the Views tab, and then double-click to open the **Composite Model View** to open it (see Figure 3–19).

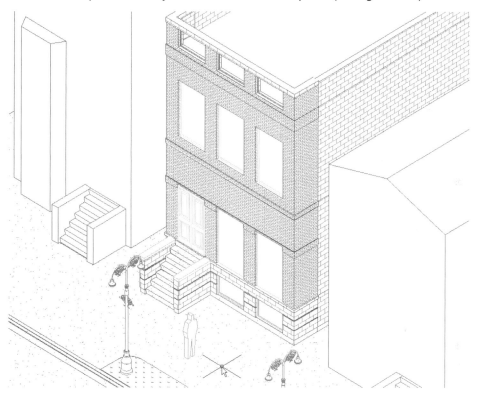

Figure 3–19 *The Composite Model View with completed configuration of custom Wall Style components*

WALL COMPONENT MATERIALS

Now that we have completed the basic dimensional setup of each component for each floor, we can now move on to assigning the correct materials to each component. A Material Definition is used in ADT to represent the actual true-life material from which a particular object will be constructed. Material Definitions control the linework and hatching for components to which they are assigned in plan, section, elevation, and 3D. Furthermore, Material Definitions (or just Materials) also can contain a render material. This render material will be visible in place of surface hatching within ADT in a shaded viewport. When we later link this model to Autodesk VIZ Render, the material assigned in ADT will remain assigned and be fully editable within VIZr.

All of the components for the façade Wall Styles have materials assigned to them already. This is because we created all of our components from existing Wall Styles. Even though we made many edits to those components, the materials assignments still remain intact. However, we need to make a few adjustments as appropriate for the design that we have devised. We will mostly use out-of-the-box Material Definitions from the library. Once again, let's start with the base-'mbment level.

Basement Level Material Definitions

Most of the Materials that we need for the rusticated base are not currently resident in the Basement Construct. For this topic, we will need to import some Materials.

1. On the Project Navigator Palette, click the Constructs tab, and then double-click to open the *Basement* Construct to open it.

 NOTE If you left the *Basement* Construct open above, then this action will simply make that file active.

2. On the View menu, choose **3D Views > SE Isometric**, and then choose **Shade>Gouraud Shaded**.

   Zoom in on the front of the building to get a better look.

As you can see, right now all of the materials are CMU or Concrete. Although this is a fine starting point, we can make the façade more interesting by introducing a bit more variety in the Material selections.

3. On the Format menu, choose **Material Definitions**.

This will call the Style Manager and filter it to show only Material Definitions. We can import Materials from an external library file and then use them in this drawing.

4. On the toolbar across the top of the Style Manager, click the **Open Drawing** icon.

   If you are not already in the *Imperial* folder, click the Content link on the Outlook Bar at left, then double-click *Styles*, and then double-click *Imperial*.

5. Locate the file named *Material Definitions (Imperial).dwg* and double-click it to open it.

6. On the left side, click the small plus (+) sign next to *Material Definitions (Imperial).dwg*, then the one next to *Multi-Purpose Objects*, and then select

Material Definitions to reveal a complete list of materials on the right that are contained in that file.

This file contains several dozen Material Definitions. The names tend to be fairly long, so you may want to resize your Style Manager as large as your monitor will allow for easy reading (see Figure 3–20).

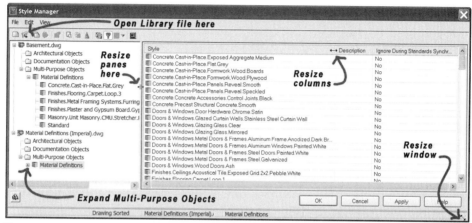

Figure 3–20 *Style Manager filtered to Material Definitions—resize the dialog and individual panes to make it more legible*

Just like Windows Explorer, select an item on the left and it will preview on the right. On the left, you can select items one at a time. On the right, you can select more than one item using the SHIFT and CTRL keys: SHIFT to select everything between the first item you click and the last item you click, and CTRL to select items in any order. To use a style from the library file in the current drawing, you must drag the style from the list at the right (or beneath the library file name on the left) to the file name of the current drawing on the left. Do *not* drag to the ADT drawing window.

7. On the right side, locate the Material Definition named **Concrete.Cast-in-Place.Panels.Reveal.Smooth**. Drag it from the right pane and drop it on top of the Material Definitions node on the left tree view pane (see Figure 3–21).

TIP You can also drop onto the Multi-Purpose folder or directly on the *Basement.dwg* node as well. If you prefer, right-click and choose **Copy** and **Paste** rather than Drag and Drop.

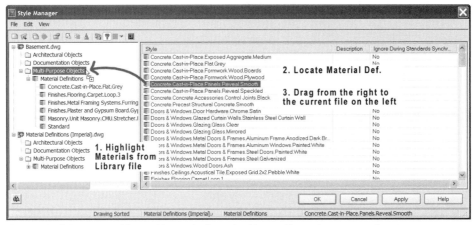

Figure 3–21 *Drag and Drop Materials from the library file to the current drawing*

For the next step, be sure that the Material Definitions node beneath the *Material Definitions (Imperial).dwg* is still highlighted.

8. Hold down the CTRL key and then click on **Finishes.Plaster and Gypsum Board.Plaster.Stucco.Fine.Brown**. Keeping the CTRL key depressed, select **Masonry.Unit Masonry.CMU.Split-Face.Running**.

9. Drag the selected Materials from the right pane and drop them on top of the Material Definitions node on the left tree view pane.

We now have the basic Materials that we will need for the basement level. However, since we have several masonry components that have similar properties but require different coursing, we will make copies of some of the Materials that we just imported and eventually edit their surface hatch parameters.

10. In the left pane, select Material Definitions beneath the *Basement.dwg* file.

11. In the right pane, right-click **Masonry.Unit Masonry.CMU.Split-Face.Running**, and choose **Copy**, and then right-click again and choose **Paste**.

ADT will append a "(2)" after the name. Accept this name.

12. Right-click and choose **Paste** again to create another copy named **Masonry.Unit Masonry.CMU.Split-Face.Running (3)**.

13. Click **OK** to close the Style Manager.

14. **Save** the file.

Assign Materials to Basement Wall Components

Assigning Materials to the Wall Style is simple. Each Component has a slot to which we can assign one of the Materials from the list. In order to assign a Material, it must first be resident in the current file. This is why we imported several Materials first.

15. Select the main horizontal Wall again, right-click, and choose **Edit Wall Style**.

16. Click the Materials tab.

You will see all of the Wall Style's components listed with a Material assigned to each one. Currently, all of the Materials assigned have come from the default designations of the original components from which we copied these. Some of these we will leave as is, others we will change. Several components: Stud, GWB, Base, and Shrinkwrap are fine the way they are. We will not change these. When you select a component in the list and then click in the Material Definition column, a dropdown menu appears from which to choose the desired Material.

17. Select the **Base Band** component and then click the Material entry next to it. From the dropdown list, choose **Finishes.Plaster and Gypsum Board.Plaster.Stucco.Fine.Brown** (see Figure 3–22).

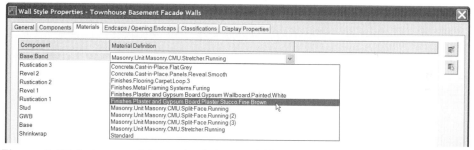

Figure 3–22 *Assign one of the previously imported Materials to the Base Band component*

18. Using the CTRL key, click to select Revel 1 and Revel 2, choose **Concrete.Cast-in-Place.Panels.Reveal.Smooth** from the Material Definition list (next to either highlighted item), and then press ENTER to accept the assignment.

19. Finally, assign **Masonry.Unit Masonry.CMU.Split-Face.Running** to Rustication 1, **Masonry.Unit Masonry.CMU.Split-Face.Running (2)** to Rustication 2, and **Masonry.Unit Masonry.CMU.Split-Face.Running (3)** to Rustication 3.

20. Click **OK** to dismiss the Wall Style Properties dialog and view the changes.

If you still have Gouraud Shading active, you will see the render textures applied to each Material. Examine the coursing of the rustication and brick components. You will notice that many of the components do not start and end on an even masonry course. This can be easily remedied by editing the hatch pattern offsets of the Material Definitions. Each of the affected components requires a different offset, which is why we have duplicated and applied several of the existing Material Definitions. Now it is just a simple matter of shifting the hatching of the pre-applied Materials.

21. Select the Wall again, right-click, and choose **Edit Wall Style** and return to the Materials tab.

Rustication 1 is fine. We will edit Rustication 2 and Rustication 3.

22. Select the **Masonry.Unit Masonry.CMU.Split-Face.Running (2)** Material next to Rustication 2.

23. In the top-right corner of the dialog, click the **Edit Material** icon.

24. In the Material Definition Properties dialog that appears, click the Display Properties tab.

25. General Medium Detail will be bold (indicating that it is active in the current viewport); select it and then click the **Edit Display Properties** icon on the top right.

26. In the Display Properties dialog, click the Hatching tab.

27. Finally, change the **Y Offset** for the Surface Hatch component to 1" (see Figure 3–23).

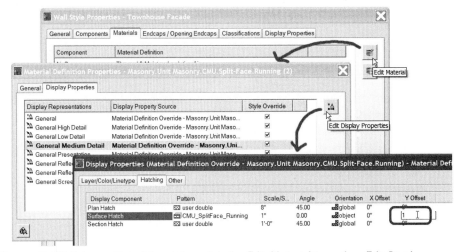

Figure 3–23 *Select a Material and then click the Edit Material icon, then Edit Display Properties to shift the Y Offset*

This will shift the surface hatch pattern of the Rustication 2 component up 1″ to compensate for the 1″ Revel component beneath it.

28. Click **OK** to exit the Display Properties dialog and then **OK** again to dismiss the Material Definition Properties dialog.

For this project, we will edit only the General Medium Detail Display Rep. If you wish to view this model at different levels of detail, you would need to repeat these steps in each of the other Display Reps.

29. Repeat this process for Rustication 3 and shift the Y Offset by **2″** this time.

30. Click **OK** to dismiss the Wall Style Properties dialog when you're finished (see Figure 3–24).

31. Choose **Hidden** from the View > Shade menu (or click the icon on the Shading toolbar).

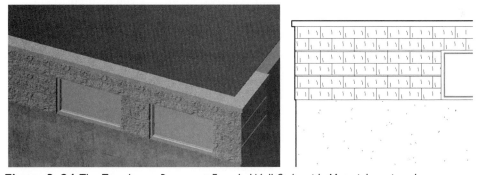

Figure 3–24 *The Townhouse Basement Façade Wall Style with Materials assigned*

32. **Save** the file.

First Floor Level Material Definitions

Following a procedure similar to the basement level, we will now import and assign Materials to the first floor.

33. On the Project Navigator Palette, click the Constructs tab, and then double-click to open the *First Floor* Construct to open it.

 NOTE If you left the *First Floor* Construct open above, then this action will simply make that file active.

34. Following a process similar to steps 3 through 7 above, import the **Masonry.Unit Masonry.Brick.Modular.Flemish** and **Finishes.Plaster and Gypsum Board.Plaster.Stucco.Fine.Brown** Materials into the current (*First Floor.dwg*) file.

35. Edit the Townhouse First Floor Facade Walls Style and assign the **Finishes.Plaster and Gypsum Board.Plaster.Stucco.Fine.Brown** Material to both the Molding 1 and Molding 2 components.

36. Assign the **Masonry.Unit Masonry.Brick.Modular.Flemish** Material to Brick 2

37. While on the Materials tab, edit the **Masonry.Unit Masonry.Brick.Modular.Flemish** Material, and set the X Offset to **–2″** and the Y Offset to **2″**.

These settings make the Flemish brick coursing match the portion of the façade upon which it is applied much better.

38. Click **OK** three times to return to the drawing and view the changes.

Our Wall Styles are shaping up nicely. As you can see, the adjustments so far have shifted all of the hatching to fall on even masonry coursing. If you wanted to cut a section through the façade wall, then you could make adjustments to the materials of the CMU components in each style to synchronize with the brick courses. However, since this component is hidden within the wall and we are concerned with rendering for this project, we can safely ignore this.

39. Following similar procedures, configure the styles in the *Second Floor* and *Parapet* Constructs to match the settings shown in Table 3–6.

 NOTE The Air Gap, CMU, Stud, and GWB components are already assigned correctly and have therefore been omitted from the table.

Table 3–6 *Townhouse Second Floor and Parapet Façade Walls Materials List*

Townhouse Second Floor Facade Walls

Component	Material Definition	Y Offset
Frieze Molding	Finishes.Plaster and Gypsum Board.Plaster.Stucco.Fine.Brown	0″
Brick	Masonry.Unit Masonry.Brick.Modular.Running	0″

Townhouse Parapet Walls

Component	Material Definition	Y Offset
Cornice Molding	Finishes.Plaster and Gypsum Board.Plaster.Stucco.Fine.Brown	0″
Brick	Masonry.Unit Masonry.Brick.Modular.Running	–2″

40. Save all files when done.

DEVELOP CUSTOM FENESTRATION STYLES

Now that we have completed the construction of our custom Wall Styles and assigned their Materials, we can turn our attention to the fenestration. The Doors and Windows that were added above were basically placeholders. We will build several Window and Door Styles to apply to the fenestration that we have already added in the next several headings.

BASEMENT WINDOWS

Let's begin with the basement windows. As we have already mentioned, it is always easier to begin with an existing style and then edit it to suit our needs. We will use the default Awning style and then apply muntins to it. ADT provides the ability to apply muntins to Windows and Doors as a display property. Muntins can be applied to the Model and Elevation Display Representations. There is even a mechanism to link the muntin display of the Model and Elevation reps together so that changes are synchronized between the two. Display Representations are controls that determine how a particular ADT object will display under separate architectural viewing situations such as Plan, Model, Elevation and Reflected. If you need more information on the Display system, refer to the online help or pick up a copy of *Mastering Autodesk Architectural Desktop* by Paul F. Aubin.

Create a New Window Style

1. On the Project Navigator Palette, click the Constructs tab, and then double-click on open the *Basement* Construct to open it.

NOTE If you left the *Basement* Construct open above, then this action will simply make that file active.

For this exercise, you can work in Front view or SE Isometric View in any shademode display. Set the display to your preferences before continuing.

2. Zoom in and select both of the basement Windows.

3. Click the Windows tab on the Tool Palettes, and then right-click the Awning tool and choose **Apply Tool Properties to Window** (see Figure 3–25).

NOTE If the Design Tool Palette Group is not active, right-click on the Tool Palettes title bar and choose Design.

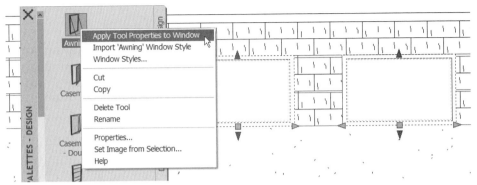

Figure 3–25 *Apply the Awning Window Tool Properties to the selected Windows*

4. With both Windows still selected, right-click, and choose **Copy Window Style and Assign**.

5. In the Window Style Properties dialog that appears, click the General tab and change the Name of the Style to **Basement Windows**.

The rest of the settings on the General tab can be configured to take advantage of the various scheduling, keynoting, and other documentation functions of ADT. None of these settings will have any impact on rendering, so we will not concern ourselves with them here. We will also skip the Dimensions tab, as all of the default settings are acceptable for the Style that we are building. Feel free to review these settings if you wish. Verify the settings on the Design Rules tab as well. Here the window Shape is set to Rectangular, and the Window Type is Single Awning. We are also not concerned with Standard Sizes or Classifications. Therefore, let's move on to Materials.

6. Click on the Materials tab.

Both the Frame and the Sash components are currently assigned to the **Doors & Windows.Metal Doors & Frames.Aluminum Windows.Painted.White Material**, while the Glass is assigned to **Doors & Windows.Glazing.Glass.Clear**. The Muntins currently have no specific assignment and use Standard. These assignments were part of the original Awning Style upon which we based our new Basement Windows Style. While the Glass Material assignment is appropriate, the Muntins ought to match the Material of the Frame and Sash.

7. From the dropdown list next to the **Muntin** component, choose **Doors & Windows.Metal Doors & Frames.Aluminum Windows.Painted.White**.

Adding Muntins

Let us now add some Muntins to our Style. In most cases, you will want muntins that you add to display in both Elevation and Model. Furthermore, it is likely that you will wish to see them in both the Model and Model High Detail Display Reps. To make this easier to accomplish, we can take advantage of an option that allows the Muntin Display of all three Display Reps to be linked together.

8. Click on the Display Properties tab.

If you are working in an isometric view, note that the Model Display Rep is bold. This indicates that it is active in the current viewport. If you are working in Front view, then the Elevation Display Rep will be bold and active instead. As indicated above, it is not important which of these two Displays is active. The muntins that we will add will be added to both, as well as Model High Detail. Currently the Property Source for all Display Reps is Drawing Default, which controls all Windows in the drawing. We need the Basement Windows to operate at the Style level (making them independent from the Drawing Default setting).

9. Next to the **Model High Detail** rep, place a checkmark in the Style Override box (see Figure 3–26).

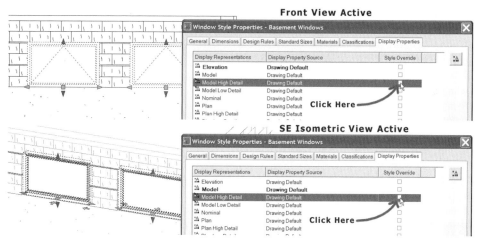

Figure 3–26 *Attach an override to the Window Style Model High Detail Rep*

10. In the Display Properties dialog that appears, click the Muntins tab, and then click the **Add** button.

11. Make no changes in the Muntins Block dialog, simply click **OK**.

12. Make certain that a checkmark appears in the **Automatically Apply to Other Display Representations and Object Overrides** checkbox, and then click **OK** (see Figure 3–27).

☑ Automatically Apply to Other Display Representations and Object Overrides

Figure 3–27 *The key to linking Elevation and Model Muntin display*

13. Back in the Window Style Properties dialog, repeat the same process for the **Model** Rep, and then click **OK** to return here again.

14. Finally, add an override and a Muntins block to the Elevation Rep as well (making certain that it also has the "automatically apply" box checked), and remain in the Muntins Block dialog this time.

Muntins blocks can be added to any pane of the Window. Obviously our Window has only one pane, so the default pane setting of Top is fine. Next you can assign the width and depth dimensions of individual muntin members. The defaults of 3/4″ wide and 1/2″ deep here are acceptable as well, but feel free to experiment with other values. With "Cleanup Joints" active (also a default), the intersections of the horizontal and vertical muntins will clean up nicely. The final default setting of "Convert to Body" is very important to this exercise. With this checked, the muntins will be made from 3D components, rather than simply drawn as lines applied to the Windows. If we were only concerned with generating a 2D elevation, this setting would not be required.

The settings in the Lights and Hub areas are used to develop your muntin pattern. There are many possible combinations. We will use a simple prairie pattern here.

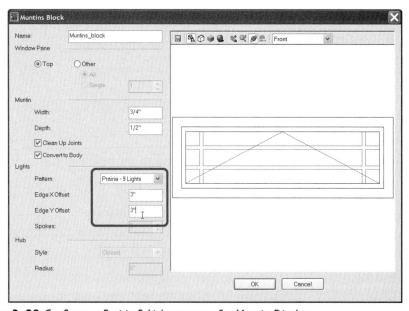

Figure 3–28 *Configure a Prairie 9 Lights pattern for Muntin Display*

15. In the Lights area, choose **Prairie 9 Lights** from the Pattern dropdown list.

16. Change both the Edge X Offset and the Edge Y Offset to **3"** (see Figure 3–28).

17. Verify the results in the embedded viewer, and then click **OK** three times to return to the drawing and view the results (see Figure 3–29).

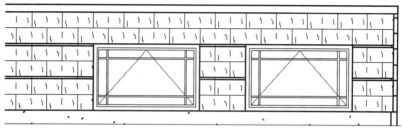

Figure 3–29 *The Basement Window style complete with muntins*

18. **Save** the file.

FIRST FLOOR DOORS

Let's move up to the first floor. Here we will get a little more complex. We are going to add half-round arches above each of the openings on the first floor. We also need to make the door on the left; which is the main entrance, a bit different than the other two.

Edit Front Door Material

We will begin the first floor modifications with the front door. Currently, the wood grain is not displaying properly on the raised panels.

1. On the Project Navigator Palette, click the Constructs tab and then double-click the *First Floor* Construct to open it.

 NOTE If you left the First Floor Construct open above, then this action will simply make that file active.

2. Change the view to SE Isometric and use the 3D Orbit to rotate the view a bit more "head-on."

 NOTE Choose Top View first before SE Isometric to be sure that the World UCS is active, or use the UCS command to make World active.

3. On the View menu, choose **Shade > Gouraud Shading** (or use the icon on the Shading toolbar).

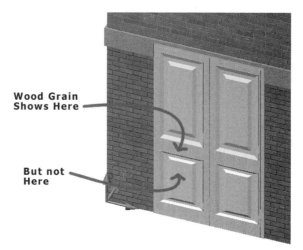

Figure 3–30 *Look carefully at the front door's raised panels; they don't display wood grain*

Notice that the wood grain texture that appears on the stiles does not show on the raised panels. This is easily remedied by making a simple modification to the Material Definition used for this door (see Figure 3–30 and Color Plate C–4a).

4. Select the front Door, right-click, and choose **Edit Door Style**.

5. Click the Materials tab (see number 1 in Figure 3–31).

6. Highlight the Frame component (which, like the Sash and Panel components, has the Doors & Windows.Wood Doors.Ash Material assigned to it), and click the small **Edit Material** icon at the top-right side of the dialog (see number 2 in Figure 3–31).

7. In the Material Definition Properties dialog, click the small **Edit Display Properties** icon (also in the top-right corner, see number 3 in Figure 3–31).

8. Finally, in the Display Properties dialog, click the Other tab, and then from the Mapping dropdown list, choose **Face Mapping** (see number 4 in Figure 3–31).

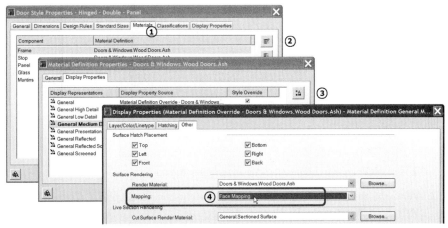

Figure 3–31 *Change the way mapping is applied to this Material Definition*

   9. Click **OK** three times to return to the drawing.

Examine the raised panels again and notice that they now are receiving the wood grain surface texture as well. (see Color Plate C–4a)

Developing a Complex Door Style

The next thing that we need to do is add a half-round arch to the top of each door. This could be done by changing the shape of the door within the Door Style; however, this approach would not create a separate pane or transom above the door, rather it would change the shape of the door itself. What we actually want is a half-round topped Window placed above each Door. The best way to do this is with a Door/Window Assembly Style.

   10. Select all three Doors on the first level.

   11. On the Design Tool Palette, right-click the Door/Window Assembly tool and choose **Apply Tool Properties to > Doors, Windows, and Openings**.

 NOTE This will override the raised panel front Door. We will reassign the raised panel style below.

   12. To be sure that all three assemblies are oriented the same way, locate the small Anchor icon on the Properties Palette and click it. Uncheck both the **Flip X** and **Flip Y** boxes and then click **OK** (see Figure 3–32).

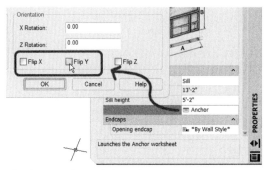

Figure 3–32 *Flip the Anchor to orient all three Assemblies the same way*

13. With the three Assemblies still selected, right-click and choose **Copy Door/ Window Assembly Style and Assign**.

14. On the General tab, rename the Style **First Floor Doors**.

15. On the Shape tab, choose **Half Round** from the Predefined list.

16. Click the Design Rules tab.

17. On the left side of the dialog, beneath Element Definitions, select **Divisions**.

This Assembly Style currently has only one Division. The Division governs the rules that determine the spacing of the mullions in the style. We need a single mullion between the top of the Door and the half-round transom.

18. On the right side, in the lower pane, change the orientation from Vertical to Horizontal. Do this by clicking the small **Horizontal** icon.

19. From the Division Type list, choose **Manual**.

20. On the far right, click the **Add Gridline** icon.

21. Change the dimensions for this gridline to **8'-1"** from **Grid Bottom** (see Figure 3–33).

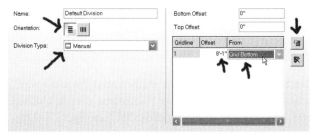

Figure 3–33 *Make necessary changes to the Default Division*

22. On the left side of the dialog, beneath Element Definitions, select **Frames**.

23. Change the size of the Default Frame to **2"** Wide by **6"** Deep.

24. Beneath Element Definitions, select **Mullions**. Repeat the same sizes as the Default Frame: **2″** Wide by **6″** Deep.

25. On the left side of the dialog, beneath Element Definitions, select **Infills**.

26. In the middle of the dialog, between the top and bottom panes on the right side, click the **New Infill** icon (see number 1 in Figure 3–34).

27. Name the New Infill **Door Infill**, and change the Infill Type to **Style**.

    Click where it says "Simple Panel" and change it to "Style" (see number 2 in Figure 3–34).

28. This will reveal a list of style types. Choose **Hinged – Double – Panel** from the Door Styles list (see number 3 in Figure 3–34).

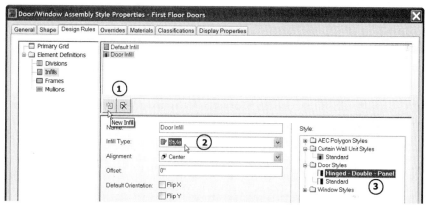

Figure 3–34 *Add a New Infill to the Style and assign the Hinged – Double – Panel Door Style to it*

29. Repeat this process to create a New Infill named: "**Transom Window Infill**," referencing the **Standard** Window Style.

30. To apply these Infill Definitions to the design, select **Primary Grid** in the left pane at the top.

The icons in the separation between the top-right and bottom-right panes will change.

31. Click the **New Cell Assignment** icon and name the new assignment **Transom**.

32. Right next to the name Transom, in the Element column, click on Default Infill and choose **Transom Window Infill** from the drop-down list.

33. Click in the Used In column. A small browse icon (…) will appear. Click this icon and uncheck **Bottom**, and then click **OK** ("Top" should be the only box checked).

34. Use the same technique to change Default Infill next to Default Cell Assignment to **Door Infill**. Leave the Used In set to All unassigned cells (see Figure 3–35).

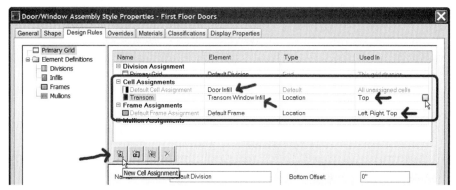

Figure 3–35 *Assign the two newly created Infill Definitions to the style*

35. Beneath Frame Assignments, click in the Used In field and then click the small Browse (...) icon that appears. Uncheck **Bottom** and then click **OK**.

36. On the Materials tab, choose **Doors & Windows.Wood Doors.Ash** for both Default Frame and Default Mullion.

37. Click **OK** to return to the drawing and then **Save** the file.

The results will not be quite as expected. There are several issues: first the height of the Doors is too short for the half-round configuration to display properly. Next, since we have referenced nested styles within the assembly, we now have two frames being expressed around three of the sides. Finally, we will need to edit the styles for the Door and Window Infills to a more desirable configuration. Let's start by increasing the height of the assemblies.

38. Click on one of the Door/Window Assembly objects.

    Be sure to click the Assembly near its frame, and not the nested Door. You can tell if the Door is selected when it highlights, because its shape will be rectangular, rather than half-round. If this happens, press ESC and then try again.

39. Click the triangular shaped grip at the top and begin moving it up.

    You will see some dynamic dimensions appear. The overall height should be highlighted in magenta. (If it is not, press TAB until it highlights.)

40. Type **10′-6″** and then press ENTER.

NOTE When you complete this action, the Assembly should refresh and the Hinged – Double – Panel Door should appear.

41. Repeat this process for the other two Assemblies.

 TIP As an alternative, you can select all three Assemblies and change the height on the Properties Palette.

First Floor Door Styles

You might have noticed that we now have three "front" doors. The middle Door and the one on the right should actually be French Doors. Let's make that edit now.

42. Select the Middle Door infill and the one on the right.

43. On the Doors Tool Palette, right-click the **Hinged – Double – Full Lite** tool and choose **Apply Tool Properties to Door**.

44. With both Doors still selected, right-click and choose **Edit Door Style**.

45. On the General tab, rename the Style **First Floor French Doors**.

46. On the Dimensions tab, set the Frame Width to **0″**.

We are doing this because these are nested within the Assembly, which has its own Frame component.

47. On the Materials tab, choose **Doors & Windows.Wood Doors.Ash** for Muntins.

48. Click the Display Properties tab, and follow the process outlined in steps 8 through 14 above under the "Adding Muntins" heading to add a Muntins Block to the same three Display Reps: Elevation, Model, and Model High Detail of this Style.

49. Use similar settings to the Basement Windows, except this time, choose **Rectangular** for Pattern, **5** Lights High, and **1** Light Wide.

50. Press **OK** three times to return to the drawing.

51. Select our one remaining front Door on the left, right-click, and choose **Edit Door Style**.

52. Change the Name to **Front Entrance Door**, and then on the Dimensions tab, change the Frame Width to **0″** and then click **OK** to return to the drawing.

First Floor Transom Window Style

Finally, let's address the transom Windows Style. The most obvious flaw is that we have a square peg being forced into a round hole (they are rectangular in shape being forced into a half-round space).

53. Select all three transom Windows. (This time we want the Window Infills, not the Assemblies, to highlight.)

54. Right-click and choose **Copy Window Style and Assign**.

55. On the General tab, change the name to **First Floor Transom**.

56. On the Dimensions tab, change the Frame Width to **0″**.

As above, we are doing this because these are nested within the Assembly, which has its own Frame component.

57. On the Design Rules tab, change the Predefined Shape to **Halfround**.

58. On the Materials tab, assign **Doors & Windows.Wood Doors.Ash** to the Frame, Sash, and Muntins components, and the **Doors & Windows.Glazing.Glass.Clear** Material to the Glass component.

59. Click the Display Properties tab, and once again follow the process outlined in steps 8 through 14 above under the "Adding Muntins" heading to add a Muntins Block to the same three Display Reps: Elevation, Model, and Model High Detail of this Style.

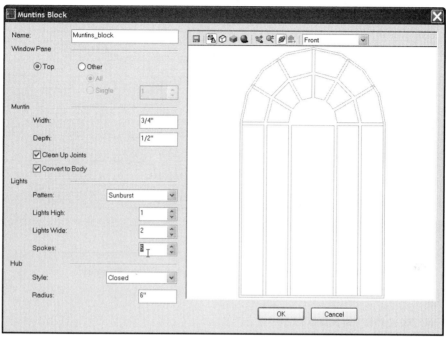

Figure 3–36 *Add a Sunburst Muntins Block to Elevation, Model, and Model High Detail Display Reps*

60. Use similar settings to the Basement Windows, except this time, choose **Sunburst** for the Pattern, with **1** Light High, **2** Lights Wide, and **5** Spokes. Accept the default of a **Closed** Hub Style with a **6″** Radius (see Figure 3–36).

61. Click **OK** three times to return to the drawing (see Figure 3–37).

Figure 3–37 *The First Floor Construct with completed Door and Window Styles*

62. **Save** the file.

SECOND FLOOR WINDOWS

The second floor windows are very similar in configuration to the first floor. They too are actually Door/Window Assemblies; we already added them as such in the "Create Second Floor Windows" heading above. Like the first level, they will consist of a larger operable window style with a transom above. Unlike the first floor, however, the transom will not be half-round, but will remain rectangular.

Create Second Floor Assembly Style

The Assemblies are already in place in the Townhouse Second Floor Facade Walls Style, so let's create the new Door/Window Assembly Style and apply the appropriate Window Style Infills within it.

1. On the Project Navigator Palette, click the Constructs tab and then double-click the *Second Floor* Construct to open it.

 NOTE If you left the *Second Floor* Construct open above, then this action will simply make that file active.

2. On the Windows Tool Palette, right-click the Casement – Double tool and choose **Import 'Casement – Double' Window Style** (see Figure 3–38).

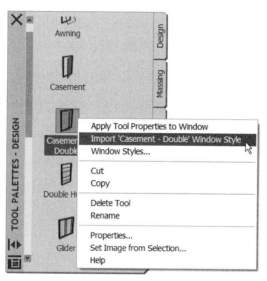

Figure 3–38 *Import an out-of-the-box Window Style to use within the Door/Window Assembly Style*

This simply imports the Window Style into the Second Floor Construct drawing and makes it available to use in this file.

3. Repeat the same process to import the **Pivot – Horizontal** Style as well.

4. To be sure that all three assemblies are oriented the same way, select all three Door/Window Assemblies and locate the small Anchor icon on the Properties Palette and click it. Uncheck both the **Flip X** and **Flip Y** boxes, and then click **OK** (see Figure 3–32).

5. With the three Assemblies still selected, right-click and choose **Copy Door/Window Assembly Style and Assign**.

6. On the General tab, rename the Assembly to **Second Floor Windows**.

7. Click the Design Rules tab. Beneath Element Definitions, choose Divisions, and highlight **Default Division**.

8. Change the orientation to **Horizontal**, the Division Type to **Manual**, and then Add a gridline **6′-0″** from **Grid Bottom**.

9. Change the dimensions of both the Default Frame and the Default Mullion to **2″** Wide by **6″** Deep.

10. Beneath Element Definitions, choose Infills, and add a new infill named **Window Infill**. Change the Infill Type to **Style**, and set the Style assignment to **Casement – Double** beneath Window Styles.

11. Add another new Infill named **Transom Window Infill** and assign the **Pivot – Horizontal** Window Style to it.

12. Repeat steps 31 through 33 above under the "Developing a Complex Door Style" heading to create and configure a Transom Cell Assignment.

13. Next, use the same technique to change the Default Infill within Default Cell Assignment to **Window Infill**. Leave the Used In set to **All unassigned cells**.

14. On the Materials tab, choose **Doors & Windows.Wood Doors.Ash** for both the Default Frame and Default Mullion.

15. Click **OK** to return to the drawing, and then **Save** the file.

Like the first floor, we have to make a few modifications to the Window Styles used as infills.

16. Select one of the transom Windows, right-click, and choose Edit Window Style.

17. On the Dimensions tab, change the Frame Width to **0″**.

18. On the Materials tab, choose **Doors & Windows.Wood Doors.Ash** for the Frame, Sash, and Muntins.

19. Click the Display Properties tab, and once again follow the process outlined in steps 8 through 14 above under the "Adding Muntins" heading to add a Muntins Block to the same three Display Reps: Elevation, Model, and Model High Detail of this Style.

20. Change the Width of the Muntin to **2″**, and change the Pattern to **Rectangular**, **1** Light High by **2** Lights Wide.

21. Click **OK** three times to return to the drawing.

22. Select all three Casement Infill Windows, right-click, and choose **Edit Window Style**.

23. Repeat steps 17 through 19 to change the Frame dimensions, assign Materials, and add the Muntins Blocks.

24. In the Muntins Block dialog, in the Window Pane area, choose **Other** and then **All** to apply the Muntins to both (all) sides of the Casement.

25. Make the Pattern **Rectangular**, **1** Light Wide by **4** Lights High.

26. Click **OK** twice to return to the drawing (see Figure 3–39).

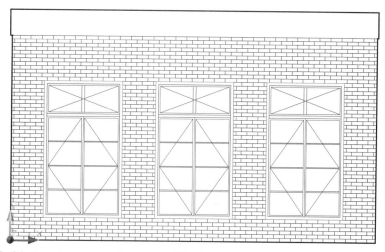

Figure 3–39 *Complete the configuration of the Second Floor Windows*

27. **Save** the file.

28. Make sure that all Constructs are saved. On the Project Navigator Palette, click the Views tab, and then double-click to open the *Composite Model* View file (see Figure 3–40).

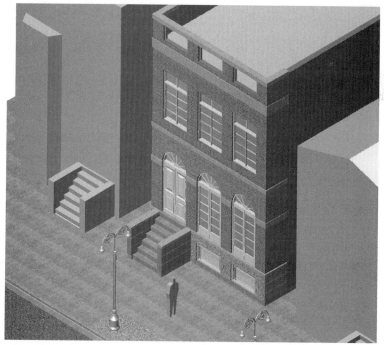

Figure 3–40 *Open the Composite Model to gauge your progress*

CONCLUSION AND SUMMARY

As you can see, we have come a long way with the façade of our townhouse. The Wall, Door, and Window Style design is complete. At this point, if you have had enough with modeling for now and would like to start using VIZ Render, skip over the next chapter and jump right into Chapter 5. In Chapter 5, "Exterior Daylight Rendering," we will use the completed Townhouse dataset and render it using a Daylight System. If you would prefer to complete the model first, then continue on to Chapter 4, where we will add all manner of accoutrement to this façade, including balconies, door hardware, arches, trim, molding profiles, and more. Looking back on what we have accomplished so far, in this chapter you learned how to:

Add Walls using Baseline Offset and Roof Line Offsets.

Work with a separated building model in Project Navigator.

Import Wall Styles from the Content Browser.

Add doors, windows, door/window assemblies, and openings.

Create and edit complex Wall Styles including "stacked" horizontal components.

Add, edit, and delete Wall Style components.

Import Material Definitions from a Style Library Content file.

Assign Materials to ADT Style components.

Edit Material Definitions including offsetting hatch patterns and modifying render material settings.

Create, import, and modify door, window, and door/window assembly Styles.

Add Muntins Display Blocks to Door and Window Styles.

CHAPTER 4

Models – Part 2

INTRODUCTION

Most of the chapters in this book will explore the various aspects of rendering and presentation from the point of view of several sample project scenarios. Before any rendering can take place, however, we must have a model from which to generate our presentations. In most of the upcoming chapters, the majority of the modeling will be provided in the datasets accompanying those chapters. In this chapter, however, we continue the task of building a model. This model will then serve as our dataset in some of the upcoming chapters. The task of building the model is a lengthy one. Therefore, the complete process is broken into two chapters. This is Part 2. The dataset for this chapter picks up where Chapter 3 left off. If you did not complete Chapter 3, and wish to start at the beginning, you need to go back and read that chapter first. Otherwise, you can begin right here.

OBJECTIVES

Although it is assumed that most readers have a basic understanding of modeling techniques in ADT, stepping beyond the basic out-of-the-box content can yield terrific results. We will explore several advanced modeling techniques in Autodesk Architectural Desktop within this chapter. Among them will be the following:

- Work in a Project Navigator project environment
- Employ advanced modeling techniques
- Work with Wall Sweeps, Endcaps, and Body Modifiers
- Work with Moldings, Trim, and Structural Members
- Work with Custom Display Blocks

THE TOWNHOUSE DATASET

The project for this chapter is a two-story townhouse. The model is provided as a series of Constructs in the ADT Project Navigator. In this chapter we will continue modeling of the townhouse façade. (Please refer to the "Models" heading in Chapter 2 for more discussion on the widely used term "Model.") In the last chapter, we built the basic Walls and Fenestration of the façade, now we will focus on some rather sophisticated accoutrement. Since this is a townhouse, the focus of this rendering project is on the front façade of the building. However, due to the nature of the ADT Project Navigator and Display System structure, we sliced this multistory façade wall into pieces; one for each floor of the project. This follows standard ADT procedures, where each floor of an ADT project is modeled in a separate file and then "stacked" up in Project Navigator to form a complete composite model. The basic structure of this model includes the following features, which will be addressed under each of the headings to follow:

Chapter 3 covered:

- ▶ Complex Wall Styles containing horizontal and vertical components used as the primary façade Walls.
- ▶ Custom Door & Window Styles: Custom settings and Muntin Display.

Chapter 4 will cover:

- ▶ Custom Door & Window Styles: Custom Endcaps and Mass Element Custom Block Display.
- ▶ Custom moldings: simple band components, Wall Sweeps, Body Modifiers and 3D Endcaps.
- ▶ Custom Structural Member Styles used as moldings.

DEVELOP BALCONY RAILING STYLES

Our façade shaped up nicely in the last chapter (take a look at Figure 3–40 to see where we left off). However, there are still many details to add before we are complete. The first thing that we want to add will be some balcony railings. We will build a single Railing Style that we will use on both the first and second floors. To make the process a bit simpler and to save a little time, two polylines have been included in the files that we will convert into railings. These polylines are on a layer that is currently frozen. We will then use the copy and assign style functionality to build a custom Railing Style.

Load the Townhouse Project

If you completed Chapter 3, you may continue in those files if you wish. The dataset for this chapter picks up exactly where Chapter 3 left off. Or feel free to work in the files provided here. If you skipped Chapter 3 and are interested in only the techniques covered here, dive right in. It is not necessary to do Chapter 3 before following Chapter 4.

1. If you have not already done so, install the dataset files located on the Mastering VIZ Render CD ROM.

Refer to "Files Included on the CD ROM" in the Preface for information on installing the sample files included on the CD.

2. Launch Autodesk Architectural Desktop from the icon on your desktop.

You can also launch ADT by choosing the appropriate item from your Windows Start menu in the Autodesk group. In Windows XP, look under "All Programs"; in Windows 2000, look under "Programs." Be sure to load Architectural Desktop, and not VIZ Render, for this exercise.

3. From the File menu, choose **Project Browser**.

4. Click to open the folder list, and choose *My Computer* and then your C: drive.

5. Browse to the *C:\MasterVIZr* folder.

6. Double-click **Townhouse04** to make this project current (you can also right-click on it and choose **Set Current Project**). Then click Close in the Project Browser.

 IMPORTANT If a message appears asking you to re-path the project, click Yes. Refer to the "Re-Pathing Projects" heading in the Preface for more information.

Create the First Floor Balcony Railing

Several Constructs have already been created for this project. There is one for each floor, including the parapet at the roof, a couple for the surrounding site information, and other miscellaneous items. Constructs are the building blocks of ADT projects. Each Construct represents a unique portion of a building model. Like a jigsaw puzzle, when all the Constructs for a project are assembled together, you have a complete building model. For more information about ADT Drawing Management, Constructs, Elements, Views, and Sheets, see the online help, or pick up a copy *Mastering Autodesk Architectural Desktop* by Paul F. Aubin. These topics are covered in detail in those references. The first task we will perform is the construction of some balcony railings.

When you loaded the Townhouse project with Project Browser above, the Project Navigator Palette should have opened automatically. If it did not, you can load it from the Window menu.

7. On the Project Navigator Palette, click the Constructs tab, and then double-click to open the *First Floor* Construct to open it.

If you are continuing from the Chapter 3 dataset, change the View to SE Isometric, and then use 3D Orbit to rotate the view slightly off the 45° angle to look at the façade more head on, but still in 3D. (The Chapter 4 dataset is already saved this way.)

8. Thaw the layer named **A-Temp**. (Use the tools on the Layer Properties Toolbar.)

A magenta Polyline will appear onscreen at the rightmost French door. If you click on the Polyline you will notice that there are several vertices. Each of these vertices will become a vertical post when the Railing is generated from it. If you wish to use this approach to create your own Railings, simply draw a Polyline with a vertex at each point where you would like a post to occur. Railing Styles can also make use of "Dynamic Posts" that are placed along the length of the Railing based on spacing parameters. Dynamic Posts work well in situations where regularly spaced Posts are desirable. In this case, the controls available for Dynamic Post placement will not give us the posts at the precise locations that we require for this design. Therefore, the Polyline vertex approach solves the problem and gives us complete control over Post placement.

9. On the Design Palette, right-click the Railing tool and choose **Apply Tool Properties to > Polyline**.

10. When prompted, select the magenta Polyline and then press ENTER (see Figure 4–1).

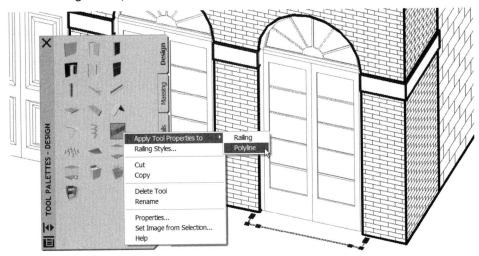

Figure 4–1 *Convert the magenta Polyline to a Railing*

11. At the "Erase layout geometry" prompt, right-click and choose **Yes**.

Create a Custom Railing Style

Leave the Railing selected and examine how each vertex of the former Polyline has been inherited as a vertex by the Railing.

12. With the Railing still selected, right-click and choose **Copy Railing Style and Assign**.

13. On the General tab, name the Style **Balcony Railings**.

14. On the Rail Locations tab, be sure that Guardrail is unchecked and that Handrail is checked.

15. Next to Handrail, set both the Horizontal Height and the Sloping Height to **2′-8″**, and the Offset from Post to **0″**.

16. Place a checkmark in **Bottomrail**, and set the Horizontal Height and the Sloping Height to **4″**.

17. In the Number of Rails text field, type **2**, and then press ENTER. (This will activate the Spacing of Rails field.) Set the Spacing of Rails value to **2′-0″**.

This will add two Bottom Rails: one at the bottom and then another 2′-0″ above it.

18. Click **OK** to see the results so far, and then **Save** the file (see Figure 4–2).

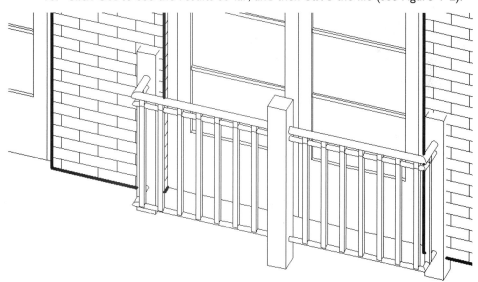

Figure 4–2 *The Balcony Railing Style with the horizontal members configured*

19. Select the Railing, right-click, and choose **Edit Railing Style**. Click the Post Locations tab.

20. Make sure that Fixed Posts is checked.

21. Change the value of both A (Extension of ALL Posts from Top Railing) and B (Extension of ALL Posts from Floor Level) to **0″**.

This will keep the posts from projecting above or below the horizontal rails.

22. Place a checkmark in the **Fixed Posts at Railing Corners** box and uncheck **Dynamic Posts**.

This will place a post only at each vertex point as discussed above (and prevent the parametrically spaced dynamic posts in between).

23. Be sure that Balusters is checked, and set D (Extension of Balusters from Floor Level) to **0″** and E (Maximum Center to Center Spacing) to **3′-0″** (see Figure 4–3).

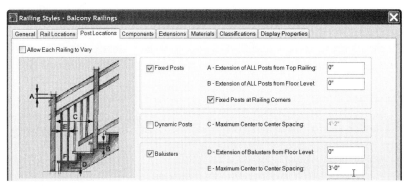

Figure 4–3 *The settings on the Post Locations tab*

For now, a single baluster will appear in the center of the Railing. Later, we will replace this baluster with a custom display block infill. See the "Railing Details" heading below.

24. Click on the Components tab.

Here we can change the shape and dimensions of all of the Railing components.

25. Change the Profile Name of all components to ***rectangular***, and the remaining settings as shown in Table 4–1.

Table 4–1 *Balcony Railing Style Component Parameters*

Component	Profile Name	Scale	Width	Depth	Rotation	Justification
A - Guardrail	*rectangular*	Scale To Fit	2″	2″	–	Middle Center
B - Handrail	*rectangular*	Scale To Fit	2″	2″	–	Middle Center
C - Bottomrail	*rectangular*	Scale To Fit	1″	1/2″	–	Middle Center
D - Fixed Post	*rectangular*	Scale To Fit	1″	1″	0	Middle Center
E - Dynamic Post	*rectangular*	Scale To Fit	1″	1″	0	Middle Center
F - Baluster	*rectangular*	Scale To Fit	1″	1″	0	Middel Center

26. Click **OK** when you have finished configuring the components.

The Railing Style is nearly complete, but let's customize the shape of the top handrail.

27. Select the Railing, right-click, and choose **Add Profile**.

28. At the "Select a railing component to add profile" prompt, click the top handrail.

29. In the Add Hand Rail Profile dialog that appears, type **Balcony Hand Rail** for the New Profile Name and then click **OK** (see Figure 4–4).

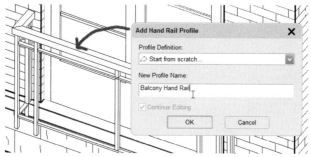

Figure 4–4 *Select the top handrail to add a new Profile Definition*

A blue shaded profile will appear cutting through the handrail. As you hover over a grip shape, a tool tip will appear to reveal its function.

Click on the blue shape. Notice the difference in grip shapes.

30. Click the small rectangular-shaped grip at the bottom edge.

31. Press the CTRL key to cycle to the Add Vertex mode. Move the mouse along the bottom edge to the left, and following the polar tracking vector, type 1/2″ (see Figure 4–5).

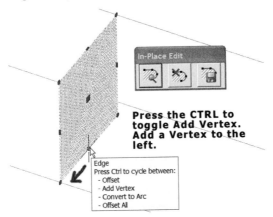

Figure 4–5 *Using the CTRL key with the rectangular grip to add a vertex*

32. Repeat the process on the other side. This time, click on the grip, press the CTRL key, and move it to the left 1/4″.

You should have two new vertices along the bottom edge; the first and last segments should be 1/2″ long and the segment in the middle should be 1″ long (see Panel A of Figure 4–6).

33. Click the edge grip (the small rectangular-shaped grip) on the newly formed 1/2″ edge at the left and drag it up. Type 1″ into the dynamic dimension field and then press ENTER (see Panel B of Figure 4–6).

This will move the edge up and create or stretch adjacent edges as required.

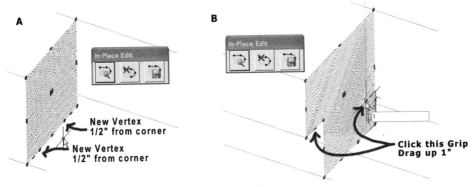

Figure 4–6 *Modify the shape of the profile by adding and stretching vertices*

34. Click the **Save all Changes** icon on the floating In-Place Edit toolbar.

The Railing Style is complete except for the custom infill block that we will add in the next topic.

35. Copy the Railing to the center French door omitting the front entrance door. (The center-to-center distance between each door is 6′-8″.)

36. **Save** the file.

CUSTOM DISPLAY BLOCKS

Much of the graphical display of ADT objects is handled parametrically by the software. Specifically, a series of Display Representations that portray the various visual aspects of a particular object are predefined within the software. Common Display Representations include Plan, Model, Reflected, and Elevation. Common components of standard objects are easy to define parametrically. For instance, the panel of a door is really just a tall thin box. Likewise a Door or Window frame is also really a box in section that is extruded along the sides of the opening as required. Sometimes, however, there are architectural components that we wish to show graphically that cannot be easily represented with the predefined display graphics. In these situations, we use Custom Display Blocks. A Custom Display Block is simply any AutoCAD block, often 3D, that is appended to a Display Rep of a particular object or style. The process of adding Display Blocks to an object or style is very similar to the method we used to add Muntins to Doors and Windows in Chapter 3. The only difference is that we must first build or import the Auto-CAD Block that we wish to apply. In this section, we will look at several examples of Custom Display Blocks. Some we will build; others we will import. Several object types in ADT can take advantage of Display Blocks. Among them are Doors, Windows, Railings, and Structural Members. Let's have a look, starting with our Railings.

RAILING DETAILS

To make our balcony railings a bit more ornate, we will add a Custom Display Block to them in place of the lone Baluster that we configured above. For this first look at custom Display Blocks, we will use a Block provided with the dataset.

Customize Railing Balusters

The Railing Infill Block utilizes several simple Mass Element Extrusions. You can try your hand at creating or modifying your own Block if you wish. (We will build a Custom Display Block from scratch below.)

1. On the Project Navigator Palette, on the Constructs tab, double-click to open the file named *Display Blocks* within the *Elements* folder (see Figure 4–7).

Contained are several pre-built AutoCAD Blocks, as well as some lines, polylines, and other geometry. We will use the sketch geometry in the coming sequence as guides for building the pieces of the first floor arch. For this exercise, we will use one of the AutoCAD Blocks to embellish our Railing Style.

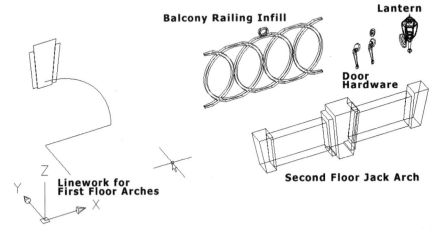

Figure 4–7 *The Display Blocks file containing several existing Blocks and linework guidelines*

2. Select the *Balcony Railing Infill* Block (see Figure 4–7), right-click, and choose **Clipboard > Copy**.

3. Switch back to the *First Floor* Construct, and then right-click and choose **Clipboard > Paste**.

TIP Try using CTRL + TAB to cycle through all currently open drawing files.

4. Don't actually place the pasted Block. Simply press ESC to cancel the command.

We only needed to import the Block Definition into the drawing. We could have placed it with the Paste command and then later erased it, but canceling the command only cancels the placement of an actual Block Reference in the drawing. Simply beginning the Paste routine is sufficient to import the Definition.

5. Select either of the Balconies, right-click, and choose **Edit Railing Style**.

6. Click the Display Properties tab, and place a checkmark in the Style Override box next to the **Model** Display Rep.

7. Click on the Other tab, and then click the **Add** button in the Custom Display Block area (see Figure 4–8).

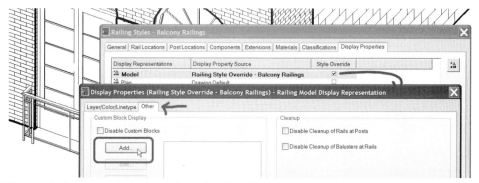

Figure 4–8 *Assign a Style Override to the Model Display Representation to add a Custom Display Block*

The Custom Block dialog will appear with lots of settings and controls that we can configure.

8. Start by clicking the **Select Block** button. From the list of Blocks that appears, choose **Balcony Railing Infill**.

There are several settings in this dialog that are unique to Railings. Many of these, however, are used when you are adding a custom Block to a Stair Railing. For instance, the Between Comp. setting, which can be used to insert a Block in between each set of posts or balusters, or the Align Slope and Shear options, which will stretch and scale the Block relative to the slope of the Stairs, keeping the vertical elements vertical. In this case, we are doing a simple balcony guardrail and only need to adjust one setting.

9. In the Component area, be sure that **Baluster** is checked, and place a checkmark in **Replace** (see Figure 4–9).

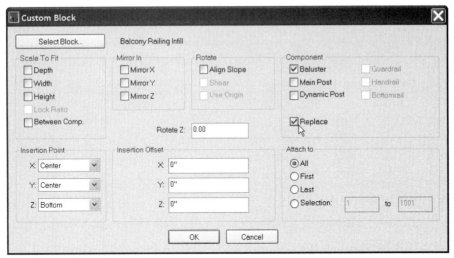

Figure 4–9 *Settings required to assign the Balcony Railing Infill Display Block to the Railing Style*

This is why we built the Railing Style above to have a single baluster in the center of the design. By choosing "replace" here, this single baluster becomes an ornate infill instead. Also, we don't have to be concerned with the "Attach To" settings, since our design has only one Baluster; attaching to "All" is the same as attaching to "First" or "Last."

10. Click **OK** twice to return to the Railing Style Properties, and then click the Materials tab.

11. Select the Guardrail component, hold down the SHIFT key, and click the Baluster component.

This will select all the components.

12. With all components thus selected, click the Material (currently Standard) next to any one of them to reveal a dropdown menu, and choose the **Metals.Ornamental Metals.Aluminum.Anodized.Dark Bronze.Satin** Material. Then press ENTER.

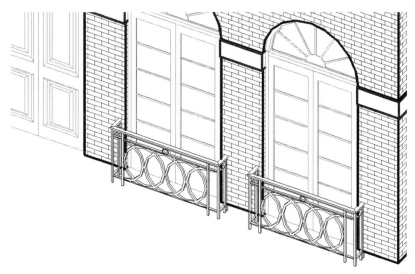

Figure 4–10 *The completed Railing Style*

13. Click **OK** to return to the drawing and then **Save** the file (see Figure 4–10).

Our Railing Style is now complete. Let's import the same Style into the second floor.

Create the Second Floor Balcony Railing

The balconies on the second floor use the same Railing Style. However, they protrude from the face of the building a bit more. Therefore, we cannot simply copy the first floor Railing objects up to the second. We must instead import the Style to the second floor and then apply it to another Polyline. To do this, we will create a custom Tool Palette and Tool.

14. On the Tool Palettes, right-click on the title bar and choose **New Palette**.

15. Name the New Palette **MasterVIZr**.

 NOTE Please be sure you have saved the file before proceeding to the next step.

16. From the Format menu, choose **Style Manager**.

17. Expand Architectural Objects and then highlight Railing Styles.

18. From the Style Manager, drag the "Balcony Railings" Style to the new MasterVIZr Tool Palette.

    A new Balcony Railings Tool should appear on the MasterVIZr Tool Palette.

19. Click **OK** to dismiss the Style Manager.

20. On the Project Navigator Palette, click the Constructs tab, and then double-click to open the *Second Floor* Construct to open it.

21. Thaw the layer named **A-Temp**. (Use the tools on the Layer Properties Toolbar.)

22. Right-click the new Balcony Railings Tool and choose **Apply Tool Properties to > Polyline**.

23. When prompted, select the magenta Polyline and then press ENTER and choose **Yes** to erase the Polyline.

24. Copy the Railing to the other two Windows. (The center-to-center distance between each window is 6'-8".)

25. **Save** the file.

FIRST FLOOR ARCHES

In Chapter 3, we created some nice French Doors on the first floor. The perfect finishing touch to these is a nice stone arch and keystone. In this example, we will both build the Custom Display Block and then apply it to the French Doors. We will use Mass Element modeling techniques to construct the arch and keystone, we will make an AutoCAD block from them, and then finally apply the block to the Window Style above the French Doors in our design.

Create the Keystone

To begin the process, we will open and work in a separate Element file.

1. On the Project Navigator Palette, on the Constructs tab, double-click to open the file named *Display Blocks* within the *Elements* folder (see Figure 4–7).

 NOTE If you left the *Display Blocks* Element open above, then this action will simply make that file active.

This is same file from which we copied the Railing Infill Block above. For this process, we will use the linework on the left to create our Mass Elements.

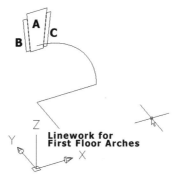

Figure 4–11 *Use the provided linework to generate the required Mass Elements*

2. Select the wedge-shaped Polyline on the left (Labeled "A" in Figure 4–11), right-click, and choose **Convert To > Mass Element**.

3. At the "Erase selected linework" prompt, right-click and choose **Yes**.

4. At the "Specify extrusion height" prompt, type **8″** and then press ENTER.

You will now have a wedge-shaped keystone Mass Element.

5. Repeat this process on the two polylines to the left and right of the keystone Mass Element that we just created (labeled "B" and "C" in Figure 4–11).

6. Erase the layout geometry again, and this time use an extrusion height of **6″**.

Now let's merge the three Mass Elements together into a single Keystone Mass Element.

7. Select one of the three Mass Elements, right-click, and choose **Boolean > Union**.

8. At the "Select objects to union" prompt, select the other two Mass Elements and then press ENTER.

9. At the "Erase selected linework" prompt, right-click and choose **Yes**.

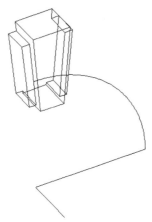

Figure 4–12 *The Keystone Mass Element result*

10. **Save** the file.

Mass Element Revolution

Mass Elements come in many predefined parametric shapes, such as box, sphere, cone, and cylinder. There are two special parametric shape types that utilize a section profile to determine their final shape: Extrusion and Revolution. And there is one additional *special* shape: Freeform. This is what the result of the union of the three Extrusions produced. A Freeform is not a parametric shape. In other words, its shape is "custom," and is derived from the result of objects from which it was made or edits applied directly to the surfaces of the Freeform shape. With Extrusion, the profile shape is "pushed" along a straight path. The keystone elements that we just completed were examples of Extrusions.

In a Revolution, the profile shape is "spun" around an axis in a full 360° circle. Revolution follows some fairly specific rules. We cannot adjust the position of the axis relative to the profile shape, nor the degrees of rotation. The axis of revolution will *always* be the X-axis of the shape used to create it, and the angle of revolution will always be a full 360° circle. These are limitations of the Mass Element Revolution. If you prefer, you can instead choose the Solid Modeling Revolve command to gain more flexibility, but later you will need to convert the solids to a Mass Element if you wish to have them respond to the ADT Display System and make use of Material Definitions. AutoCAD Solids, although powerful, can use neither of these ADT features directly. Therefore, for this exercise, and despite its limitations, we will explore the Mass Element Revolution.

 NOTE If you choose to use REVOLVE instead, be sure to use the Convert to Mass Element tool on the Massing Tool Palette to convert your solid to a Mass Element before continuing on with the exercise.

Let's now create a Revolution Mass Element that will become the basis of the Arch. Revolution Mass Elements are tricky to master. Unlike Extrusion Mass Elements, which can be created by directly right-clicking a Polyline, to create a Revolution Mass Element, you first need an AEC Profile Definition (or simply "Profile"). A Profile is basically a named closed Polyline. You can create one by drawing a closed Polyline, right-clicking it, and using the "Convert to Profile" option on the right-click menu. Next, you create a Revolution Mass Element from the Massing Tool Palette and select the Profile for it from the Properties Palette. The way in which the Profile is revolved is controlled by very specific (hard-coded) parameters. Take a look at Figure 4–13. This illustration demonstrates the way in which an ADT Profile shape is interpreted and utilized by the Revolution Mass Element. Draw your closed Polyline anywhere in the drawing. The trick to getting the desired result in the Revolution is to realize that the X-axis of your shape will become the axis of revolution, and the Y-axis will become the "base" of the object, or in other words, will define a plane parallel to the current UCS (User Coordinate System).

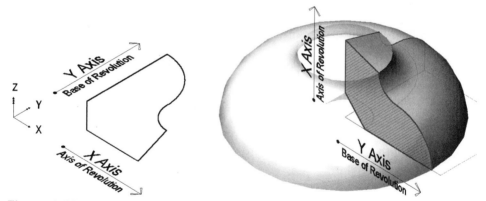

Figure 4–13 *Understanding how a Profile shape is used by the Revolution Mass Element*

Look at the UCS icon onscreen (an example is visible in the lower-left corner of Figure 4–11). It indicates which direction is the positive X, Y, and Z direction in the drawing, and serves as a type of compass as you work.

    11. From the View menu, choose **3D Views > Top**. (Or use the Top view icon on the Views toolbar.)

Notice that the X-axis now points directly to the right, while Y is pointing directly up. This is the default "World" coordinate system (WCS). The preset orthographic view commands in ADT typically also change the orientation of the UCS to make it parallel to the screen.

TIP Right-click on any toolbar icon and choose UCS II from the menu. This will load the UCS II toolbar, which provides several very useful predefined User Coordinate Systems just a click away.

12. On the shapes toolbar, click the Polyline icon.

13. Click the first point anywhere in a blank area of the screen. Move the mouse directly to the right, and using Polar (or Ortho) to keep the line level, type **4″** and then press ENTER.

14. Next, move the mouse directly up (90°) again with Polar or Ortho, and type **3′-1″** and then press ENTER.

For the next point, we will key in a relative coordinate in standard AutoCAD format.

15. At the command line, type **@-1″, 1″** and then press ENTER.

This will create a 45° diagonal line up and to the left. If you are unsure about keying in AutoCAD coordinates, consult the online help.

16. Continue in a similar fashion, moving next to the left (180°) and typing **3″** (see left panel of Figure 4–14).

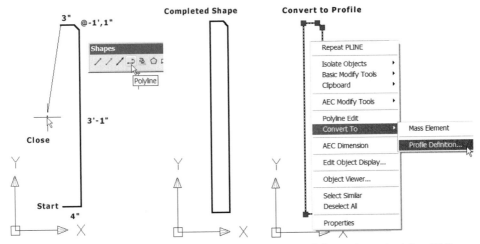

Figure 4–14 *Draw the segments of the polyline shape using Polar and standard AutoCAD coordinate entries*

17. Finish the polyline by right-clicking and choosing **Close**. (You can also type **C** and then press ENTER.) The completed shape is shown in the middle of Figure 4–14.

18. Select the polyline, right-click, and choose **Convert To > Profile Definition** (see the right panel of Figure 4–14).

19. At the "Insertion point" prompt, right-click and choose **Centroid**.

20. At the "Profile Definition" prompt, right-click and choose **New**.

21. In the New Profile Definition dialog, type **Arch Profile** for the name and then click **OK**.

22. **Save** the file.

23. From the View menu, choose **3D Views > Front.** (Or use the Front view icon on the Views toolbar.)

On the left side of the screen, you will be looking straight at the keystone we created above. In close proximity is a line and an arc. We will use these as a guide to place the new Revolution object.

24. On the Massing Tool Palette, click the **Revolution** tool.

25. On the Properties palette, in the General grouping, choose **Arch Profile** from the Profile list.

26. At the "Insert Point" prompt, click the endpoint of the line that is directly beneath the keystone Mass Element (see Figure 4–15).

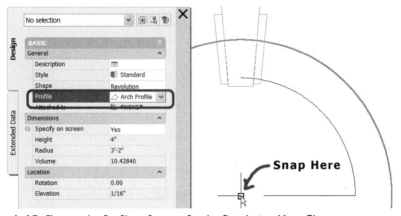

Figure 4–15 *Change the Profile reference for the Revolution Mass Element*

27. At the "Rotation" prompt, press ENTER to accept the default rotation of 0°, and then press ENTER again to complete the command.

We now have a Revolution Mass Element that conforms to the size of the Profile drawn above.

28. On the Massing Tool Palette, click the **Cylinder** tool. At the "Insert Point" prompt, click the same endpoint (of the line that is directly beneath the keystone Mass Element) again.

29. At the "Radius" prompt, click the other endpoint of the line.

30. At the "Height" prompt, type **12″** and press ENTER. Finally, accept the default rotation by pressing ENTER a second time. Press ENTER one more time to complete the command.

31. From the View menu, choose **3D Views > SW Isometric**. (Or use the SW Isometric view icon on the Views toolbar.)

32. Select the Revolution Mass Element, right-click and choose **Trim by Plane**.

A red outline will appear around the Revolution.

33. At the "Specify trim plane start point" prompt, click on the same endpoint of the line that we centered the two Mass Elements upon (see Panel A of Figure 4–16).

34. At the "Specify trim plane end point" prompt, move the cursor horizontally, maintaining a 0° angle and then click again (see Panel B of Figure 4–16).

35. Finally, move the cursor up and down and notice that a red highlighting will appear around half of the Mass Element. Move the cursor down, and click to remove the bottom half in answer to the "Select mass element side to remove" prompt (see Panel C of Figure 4–16).

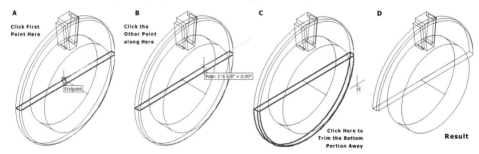

Figure 4–16 *Follow the prompts of the Mass Element Trim command to remove the bottom half of the Revolution*

36. Select the now trimmed Revolution Mass Element, right-click, and choose **Boolean > Subtract**.

37. At the "Select objects to subtract" prompt, select the cylinder and then press ENTER. Answer **Yes** to the "Erase layout geometry" prompt.

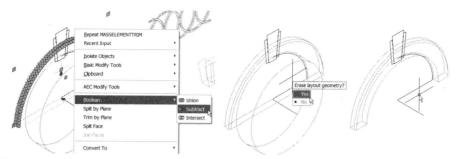

Figure 4–17 *Subtract the cylinder from the revolution*

38. **Save** the file.

Create the Half-Round Arch Display Block

Although we could now Boolean the arch and the keystone together, we will leave them separated because it will ultimately give us greater flexibility. For instance, with the two Mass Elements remaining separate, we can later choose to assign them different materials if we wish. For now we will use the same material, but later in Chapter 5 we will explore different material options.

39. Select both Mass Elements (the keystone and the arch), and on the Properties palette, choose **Arch** from the Style list.

This Mass Element Style was already resident in the Display Blocks Construct. It assigns the proper Material Definition to the Mass Elements.

40. On the UCS II toolbar, choose **World** from the dropdown list (see Figure 4–18).

TIP You can also type UCS at the command line, press ENTER, and then type **W** and then press ENTER again. See Chapter 2 for more information.

Figure 4–18 *Change the UCS back to World before creating the Block*

41. With the two Elements still selected, choose **Blocks > Block Definition** from the Format menu. (Or simply type **B** at the command line and press ENTER.)

42. For the Name, type **Half Round Arch**. Verify that the Base Point is set to 0,0,0 and that the Object reads, "2 Objects Selected," and then click **OK** (see Figure 4–19).

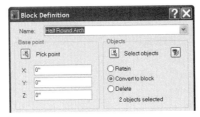

Figure 4–19 *Verify the Block Definition settings*

 NOTE The endpoint of the line that lay beneath the keystone is at the origin point – 0,0,0. This was done for convenience.

43. **Save** the *Display Blocks* file. (Don't close the file.)

The object will appear to "blink" and will be converted directly into a Block. You can verify this if you wish by clicking on it. Notice that it now has a single grip at the insertion point. We now have completed our block. And it is ready to be added to the First Floor Windows.

There are several other blocks already in this file. There is a Jack Arch Block that was created completely from simple Extruded Mass Elements, there are two door hardware Blocks and an outdoor lantern Block that are comprised of several free-form Mass Elements, and finally there is the Balcony Railing Infill Block that we have applied to our Railing Style already. It uses Mass Element Extrusions and Boolean operations to subtract out the inner voids. Let's apply these Blocks now to our townhouse façade.

Apply Custom Display Blocks

Let's move to the First Floor Construct, and add the Block that we just created to the first floor half-round arches. In Chapter 3 the Half-Round French Doors were created for the first floor. These are actually Door/Window Assembly objects in which a Double Door Style is assigned to the bottom portion and a half-round transom Window Style is applied above. The Display Block that we just completed will be assigned to the transom Window Style.

44. Select the two Door Hardware Blocks (*Handleset* and *Handleset Deadbolt*), the *Lantern* Block (see Figure 4–7), and the *Half-Round Arch* Block that you just created, right-click, and choose **Clipboard > Copy**.

45. On the Project Navigator Palette, on the Constructs tab, double-click to open the file named *First Floor* within the *Constructs* folder.

NOTE If you left the *First Floor* Construct open above, then this action will simply make that file active.

46. Switch back to the *First Floor* Construct, and then right-click and choose **Clipboard > Paste**.

47. Don't actually place the pasted Blocks. Simply press ESC to cancel the command.

As noted above, this is all that is needed to import the Block Definitions into the First Floor file. If you prefer to see them as you work, go ahead and insert them into the file off to the side. Later when we have finished adding all of the Blocks, you can erase them.

48. Select one of the first floor half-round transom windows, right-click, and choose **Edit Window Style**.

49. Click the Display Properties tab, highlight Model and then click the **Edit Display Properties** icon (see Figure 4–20).

Figure 4–20 *Edit the Properties of the existing Model Style Override*

50. Click the Other tab, and then click the **Add** button in the Custom Block Display area.

51. Click the Select Block button and choose *Half Round Arch* and then click **OK**.

If you look in the embedded Viewer, it appears as though the Display Block is too large. This is because the Viewer is showing a default graphic for the Window Style and the default width of a Window is only 3′-0″, where our Window is a bit wider at 4′-4″ (minus the 2″ frame of the Window Assembly surrounding it). Don't be concerned about this. Even though the viewer is misleading, the result in the drawing will be correct.

52. Place a checkmark in the **Mirror Y** checkbox.

The default orientation of the Block would place it on the inside of the Wall. This checkbox will flip the arch to the outside of the building.

53. In the Insertion Point area, set the **X** and **Y** values to **Center**.

Because we must mirror the Block, and because the insertion point of the Block is on the edge of the Block, we also have to apply an offset to the Block Insertion. To calculate the proper offset, we have to consider the overall size of the Block (which is 8″ in the Y direction) and the distance from the center (since we chose Center for the Y Insertion Point) of the Window to the face of the Wall (which is currently 5 1/4″). However, since we are mirroring the Block, we must treat the Block's Y dimension as a negative. Therefore, we need to add –8″ and 5 1/4″, for a result of –2 3/4″. This is the value we need to use in the Y Insertion Offset.

54. In the Insertion Offset area, type **–2 3/4″** for Y, and then click **OK** twice to return to the Display Properties tab of the Window Style Properties dialog.

Unfortunately, Display Blocks do not have the automatic synchronization setting like Muntins (as seen in Chapter 3), so therefore we have to add the Display Block to each Display Rep where it is required.

55. Repeat these exact steps on the **Elevation** and **Model High Detail** Display Reps.

56. Click **OK** to return to the drawing (see Figure 4–21).

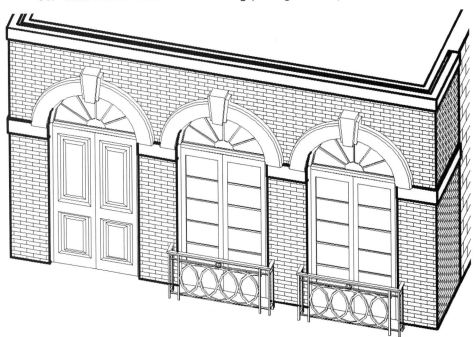

Figure 4–21 *Adding the Half-Round Arch Custom Display Block to the First Floor Transom Windows*

57. **Save** the file.

SECOND FLOOR ARCHES

The Block for the second floor was created very much like the block for the first floor. Each of its components are simply extruded Mass Elements. We can perform nearly the same steps to add the Jack Arches to the transom windows on the second floor.

1. On the Project Navigator Palette, on the Constructs tab, double-click to open the file named *Display Blocks* within the *Elements* folder.

NOTE If you left the *Display Blocks* Element open above, then this action will simply make that file active.

2. Select the Second Floor *Jack Arch* Block (see Figure 4–7), right-click, and choose **Clipboard > Copy**.

3. On the Project Navigator Palette, on the Constructs tab, double-click to open the file named *Second Floor* within the *Constructs* folder.

NOTE If you left the *Second Floor* Construct open above, then this action will simply make that file active.

4. Right-click and choose **Clipboard > Paste**. As before, don't actually place the pasted Block, simply press ESC to cancel the command.

The Windows on the second floor swing out. The balcony Railings that we built for them above are deeper to facilitate this, however, the Windows themselves should be shifted back within the Wall a bit to allow for better clearance. You can verify this by selecting one or more of the nested Windows, and changing the Opening Percent on the Properties Palette to 90°. Notice how close the Window Sashes come to the Railings.

5. From the View menu, choose **3D Views > Top** (or use the Top view icon on the Views toolbar).

6. Select all three Door/Window Assemblies. (Make sure you highlight the Assemblies and not the nested Window infills. See the left side of Figure 4–22.)

TIP To select just the Assemblies, click on the left or right side Frame immediately adjacent to the Wall.

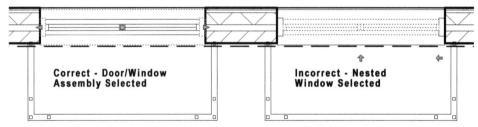

Figure 4–22 *Select only the three Door/Window Assemblies, not the Nested Windows*

Look carefully at the objects selected; the Door/Window Assemblies contain the sill lines and the Frames at the left and right sides of the opening (see the left side of Figure 4–22). The nested Window highlights the infill only and the Window Sashes, but not the Frames (see the right side of Figure 4–22).

7. Right-click and choose **Reposition Within Wall**.

8. At the "Select position on the opening" prompt, move the red line to the top edge of the Assembly and click (see panel A in Figure 4–23).

9. At the "Select a reference point" prompt, click the Endpoint directly above the right corner of the Assembly on the Wall (see panel B in Figure 4–23).

10. Finally, at the "Enter the new distance between the selected points" prompt, type **0″** and then press ENTER (ssee panel C in Figure 4–23).

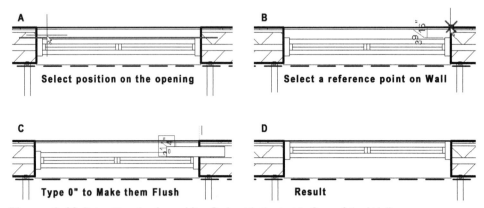

Figure 4–23 *Reposition the Assemblies flush with the inside face of the Wall*

If you changed the Opening Percent of one of the Windows, you can now see that the new position within the Wall gives use better clearance with the Railing. Go ahead and set the Opening Percent back to 0°.

11. From the Views menu, choose **3D Views > SE Isometric** (or use the SE Isometric view icon on the Views toolbar).

12. Select one of the second floor transom windows, right-click, and choose **Edit Window Style**.

13. On the Display Properties tab, highlight Model, click the Edit Display Properties icon, and then click the Other tab.

14. Click the **Add** button in the Custom Block Display area.

15. Click the **Select Block** button, choose *Jack Arch*, and then click **OK**.

16. Place a checkmark in the **Mirror Y** checkbox.

17. In the Insertion Point area, set the **X** and **Y** values to **Center** and choose **Top** for Z.

Unlike the Half-Round Arch Block used in the first floor Construct above, this Block must also shift in the Z direction. The "Top" choice moves the Block to the top of the Window. However, like the Block on the first floor, we again have to apply an offset to the Block Insertion to compensate for the combination of the Window's position within the Wall and the Block's Base Point. The overall size of this Block is also 8″ in the Y direction, however this time the distance from the center (since we chose Center for the Y Insertion Point) of the Window to the face of the Wall is now 8 1/2″ (after we shifted its position above). However, as with the first floor arch Block above, since we are mirroring the Block, we must again treat the Block's Y dimension as a negative. Therefore, we need to add –8″ and 8 1/2″, for a result of 1/2″ this time. This is the value we need to use in the Y Insertion Offset.

18. In the Insertion Offset area, type **1/2″** for Y and then click **OK** twice to return to the Display Properties tab of the Window Style Properties dialog.

19. Repeat these exact steps on the **Elevation** and **Model High Detail** Display Reps.

20. Click **OK** to return to the drawing (see Figure 4–24).

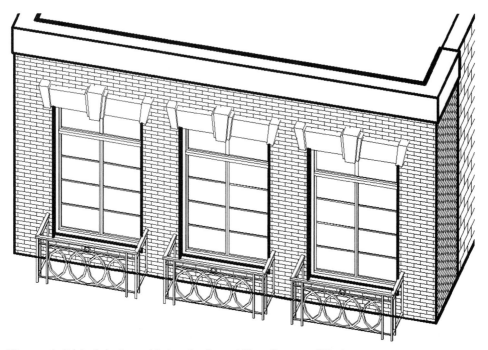

Figure 4–24 *Jack Arches added to the Second Floor Transom Windows*

    21. **Save** the file.

FIRST FLOOR DOOR HARDWARE

We'll now return to the first floor and add hardware to the front door using a similar process. The Display Blocks that we need were already imported into this file in steps 44 through 46 of the "Apply Custom Display Blocks" heading above. Let's apply them to the Front Door now.

Add Door Handles

    1. On the Project Navigator Palette, on the Constructs tab, double-click to open the file named *First Floor* within the *Constructs* folder.

 NOTE If you left the *First Floor* Construct open above, then this action will simply make that file active.

    2. Select the Front Entrance Door (the one with the raised panels), right-click, and choose **Edit Door Style**.

    3. As we did for the Windows above, click the Display Properties tab, select Model, and then click the Properties icon.

TIP You can also double-click on Model to open Properties.

4. Click the Other tab, and then click the **Add** button to add a Custom Display Block.

NOTE This Door Style already has a Display Block that is used to represent the raised panel.

5. In the Custom Block dialog, click the Select Block button and choose *Handleset Deadbolt.*

Notice that the block appears at the bottom-left corner of the Door in the viewer. We need to adjust some settings to get it to appear in the correct location. Start in the bottom of the dialog in the Component area. This is where we tell the block to which part of the Door it belongs. Since this is the door handle that we are adding, it should belong to the Door Leaf Component. In this way, if the leaf changes, the door hardware will follow. For instance, if you were to open the door, the handle would stay attached and follow the relative position of the door leaf. When you choose either the Leaf or Glass component, you also must decide if it should apply to all such components in the style or just a particular one. If you choose a single one, you must then type a number in the index field to indicate which one. That is what we will do here, since we have a different Block for each leaf of the double door. This can sometimes be guesswork. But in this case, since there are only two leafs, it matters little which is "1" and which is "2."

6. In the Component area, choose **Leaf Component** and then **Single**. Type I in the field immediately adjacent (see item I in Figure 4–25).

7. Next, change the X Insertion Point to **Left**, the Y to **Back**, and leave the Z set to **Bottom** (see item 2 in Figure 4–25).

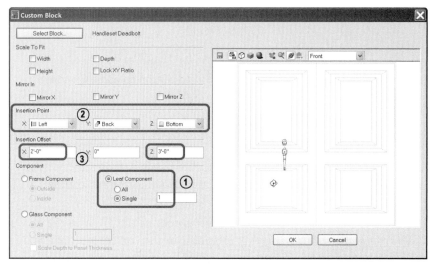

Figure 4–25 *The settings required to add the Door Hardware Custom Block*

The key to success here is being sure that the block has been defined with a logical base point and in a useful orientation. As you can see, there are controls to mirror a block, but not to rotate it. Also, when you mirror, the insertion point and the overall extents of the block influence the final position (like we saw above). The last thing that we need to do is shift the block in the X and Z directions to position it in the correct location relative to the Door Leaf.

8. In the Insertion Offset area, change the X Offset to **2′-0″** and the Z Offset to **3′-0″** (see item 3 in Figure 4–25).

9. Double-check your settings with Figure 4–25 and then click **OK** three times to return to the drawing.

You will notice that there can be a bit of a disparity between the image shown in the embedded viewer and the actual result in the drawing. This is a good reason to OK out of the dialogs every once in a while to see your progress. This also gives you a good opportunity to save your file.

10. Repeat the process to add the *Handleset* Block to the other leaf.

This time, the X Insertion Point is **Right**, the X Offset is **-2′-0″**, and the Leaf Component is **2** (see Figure 4–26).

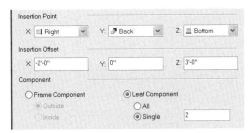

Figure 4–26 *Add the Handleset Block to the other Door Leaf*

11. OK back to the Display Properties tab of the Door Style Properties dialog, and repeat this process again to add both Blocks to the **Elevation** and **Model High Detail** reps. Use all the same settings for both reps.

12. **Save** the Drawing.

Edit Door Profile

If you zoom in on the door and look carefully at what we have so far, you will notice that the door handle overlaps the raised panel. The reason for this is that the raised panels are also Display Blocks; however, they have been assigned to the Glass component. Since the raised panel block completely fills the glass component, we get the appearance of a raised panel door. If you edit the Door Style again, and edit the AEC_Raised_Panel Display Block, you can see this for yourself. Upon doing so, you will further notice that the AEC_Raised_Panel Block is set to scale in the X, Y, and Z dimensions. This is what makes it "fill" the Glass components. Therefore, what we need to do is to increase the size of the door's stiles (thereby decreasing the width of the glass components and their associated raised panels). To do this, we must edit the Profile that gives shape to the "Glass." This can be done directly in the drawing. (We have already worked a bit with Profiles when creating the Revolution Mass Element and when customizing the Railing.) If you are in the Door Style dialog, be sure to click OK to return to the drawing before continuing.

13. Select the Front Door, right-click, and choose **Edit Profile in Place**. Click **Yes** in the alert that appears regarding size and scale.

We want the middle door stiles to be wider (making the panels narrower).

14. To do this, use the grips on the Edit in Place Profile and Polar or Ortho to add 1″ in each direction.

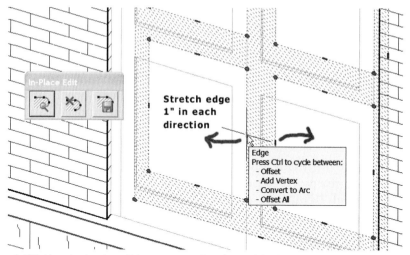

Figure 4–27 *Use the In-place Edit grips to widen the middle Door Stile and simultaneously narrow each of the panels*

15. Repeat this four times; once for each panel left and right, top and bottom. When finished, click the **Save All Changes** icon on the In-Place Edit toolbar (see Figure 4–27).

16. **Save** the file when finished.

If you wish, feel free to add the same hardware to the French Doors. Or if you wish, you can try your hand at creating your own hardware block to add. If you plan to build your own, you may want to do this exercise first. For now, we will leave the French Doors without hardware and move on to our final Display Block exercise.

Add Lanterns

As a final finishing touch to the first floor, let's insert the last remaining Block (the Lantern Block) that we imported previously. This Block will not be applied as a Display Block, but rather will simply be inserted on the façade in the correct location.

17. From the Insert menu, choose **Block** (or simply type **I** and then press ENTER).

18. From the Name list, choose *Lantern*, place a checkmark in the "Specify On-Screen" checkbox for Insertion point, and do *not* check it for Scale and Rotation.

19. Make sure that Explode is *not* checked, and then click **OK**.

20. At the "Specify insertion point" prompt, snap to the Midpoint of the band molding (Molding 1) next to the front Door (see Figure 4–28).

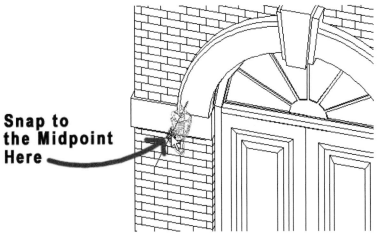

Snap to the Midpoint Here

Figure 4–28 *Insert Lantern at the Midpoint of the Band Molding 1*

21. Insert a second one on the other side of the Door, also at the Midpoint of the band molding.

22. Select both of the *Lantern* blocks in the drawing, and on the Properties Palette, change the Position Y to **0"**, the Position Z to **6'-8"** and the Rotation to **270°** (see Figure 4–29).

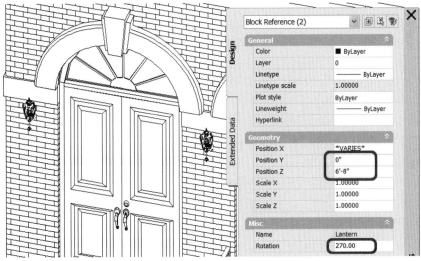

Figure 4–29 *Shift and Rotate both Lanterns into proper position and orientation*

23. **Save** the file.

TRIM AND MOLDINGS USING SWEEPS

In this section, we will explore various techniques to apply trim and moldings to our architectural façade. Although this is an exterior model, many of these techniques can be used on interior models as well. There are basically two approaches to trim: make it integral to the Wall Style or apply it as a second object. There are advantages and disadvantages to both approaches. When you make it integral to the Wall Style, you have two options. For simple band and board moldings, you can define a Wall Style Component as we have done in the Molding 1 and Molding 2 components (see Chapter 3). If you wish to make the profile of these band moldings anything other than rectangular, you must then apply a Sweep. A Sweep utilizes a Profile (like the one we defined above for the Revolution Mass Element), and "sweeps" or extrudes it along a component of the Wall. The technique is very effective and gives a nice effect. You can start by defining a profile and then applying that to the Sweep, but it is often easier to simply use the In-Place Edit functionality to create the Sweep directly on the model. Sweeps do have some limitations as we will see below. When you wish to achieve techniques not possible with Sweeps, you can apply your trim as a separate object instead. Let's look first at Sweeps, and then we will discus applied moldings below.

Sweeps via In-Place Edit

In this heading, we will start with an In-Place Edit Sweep applied to the lowest band molding component in our façade Wall.

1. On the Project Navigator Palette, on the Constructs tab, double-click to open the file named *Basement* within the *Constructs* folder.

   If you continued from your Chapter 3 dataset, use the SE Isometric view as a starting point, and then use 3D Orbit to rotate the view slightly so that we are facing the front wall more "head-on."

2. Select the three façade Walls. (The main front façade Wall and the two short Walls immediately adjacent to it.)

3. Use your wheel (or any zoom method that does not disrupt the object selection) to Zoom in on the Base Band molding at the top of the façade Wall.

4. With the three Walls still selected, right-click and choose **Sweeps > Add**.

5. In the Add Wall Sweep dialog, choose **Base Band** from the Component list.

6. From the Profile Definition list be sure that Start from scratch is chosen, and type **Façade Wall Base Band** for the New Profile Name.

7. Be sure that there is a checkmark in the **Miter Selected Walls** and the **Apply Roof/Floor Lines to Sweeps** checkboxes, and then click **OK** (see Figure 4–30).

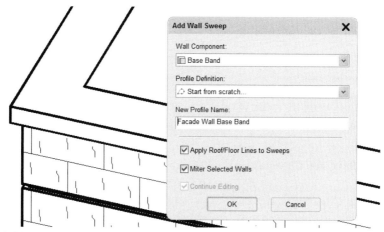

Figure 4–30 *Create a New Sweep Profile from scratch for the Base Band component*

8. At the "Select a location on wall for editing" prompt, click a point on the Base Band component anywhere on the main horizontal façade Wall.

A cyan-colored shaded profile shape with grips will appear at the location you indicated. You can edit this shape directly and see the result on the Base Band component in real time. The grips can be used to edit the shape as well as a variety of right-click options. You can also draw a Polyline, and then right-click to replace the default profile shape with the shape of the Polyline (using Replace Ring). In this exercise, we will make a very simple modification to the provided shape. We want to edit the vertical edge of the Base Band component that is facing us to give it a beveled effect.

Hover over the Edge Grip and notice that there are three toggle modes available using the ctrl key. These modes are Offset, Add Vertex, and Convert to Arc.

9. Click the Edge grip and begin stretching it. Press the CTRL key twice.

This will cycle to the Convert to Arc option.

10. Pull the grip out away from the face of the Wall using Polar or Ortho, type **3/4"**, and then press ENTER (see Figure 4–31).

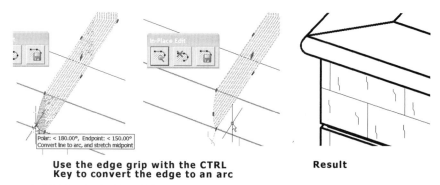

**Use the edge grip with the CTRL
Key to convert the edge to an arc**   **Result**

Figure 4–31 *Use the In-Place Edit to create a beveled profile on the Base Band Wall Component*

11. Click **Save All Changes** on the In-Place Edit Toolbar

If your result does not look smooth, type FACETDEV at the command line and then press ENTER. Type .01 and press ENTER again. REGEN if necessary. FACETDEV determines the smoothness of ADT objects. The lower you set the number, the smoother your curved shapes will be. However, if you go too smooth, it can adversely affect performance.

Let's shift our focus to the top of the façade. We will open the Parapet Construct and perform another example.

12. On the Project Navigator Palette, on the Constructs tab, double-click to open the file named *Parapet* within the *Constructs* folder.

13. **Thaw** the Layer named **A-Wall-Adtv**.

You should see a magenta Polyline appear at the top of the façade.

14. Select all three façade Walls (like we did in the *Basement* file above) and then Zoom in on the Polyline.

In this sequence, you can use the In-Place Edit profile to sculpt the Cornice Molding to the shape of the Polyline. If you were doing this process on an actual project, you would not be required to draw a Polyline first. The Polyline was provided here merely for convenience. If you prefer working with Polylines, however, you can also simply right-click and then choose Replace Ring and select the Polyline provided. We will use that technique in the next exercise. In this sequence, we will go through the process with the In-Place Edit. Feel free to use the "Replace Ring" option if you prefer.

15. With the three Walls selected, right-click and choose **Sweeps > Add**.

16. In the Add Wall Sweep dialog, from the Wall Component list, choose **Cornice Molding**. Leave Profile Definition set to **Start from scratch**.

17. For the New Profile Name, type **Cornice Molding Profile**, making sure that there is a checkmark in the **Miter Selected Walls** and the **Apply Roof/Floor Lines to Sweeps** checkboxes, and then click **OK**.

18. At the "Select a location on wall for editing" prompt, click a point near the Polyline.

TIP Don't click directly on the polyline, but close to it. You have to click on the Wall for it to accept your pick.

19. Click on the cyan-colored In-Place Profile to reveal its grips. Click on the edge Grip facing the front of the façade.

20. Press the CTRL key to toggle to Add Vertex mode.

21. Snap the new vertex to the vertex of the Polyline closest to the top of the profile (see the left panel of Figure 4–32).

NOTE The closer you picked for the "location on wall for editing" prompt, the easier this will be; however, you can still snap to the Polyline points even if the In-Place Edit Profile is not immediately adjacent. You may also move the Polyline closer to the In-Place Edit Profile if you wish.

22. Repeat this process to add vertices and snap them to the shape of the Polyline at all points. Use Figure 4–32 for guidance.

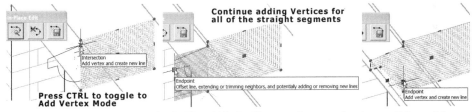

Figure 4–32 *Add vertices to the In-Place Edit Profile to sculpt it to the shape of the polyline*

23. To create the arc segment, add all other vertices until there is only one segment remaining connecting the two endpoints of the arc (left panel of Figure 4–32).

24. Press the CTRL key twice to cycle to the "convert to arc" mode.

25. Stretch the point to the midpoint of the arc as shown in the middle of Figure 4–33.

26. Click the **Save All Changes** icon on the In-Place Edit toolbar to complete the Sweep.

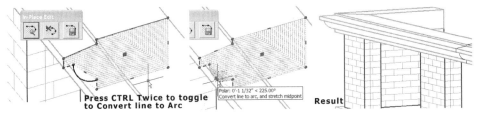

Figure 4–33 *Finalize the shape by adding the arc segment*

27. Delete the magenta Polyline and then **Save** the file.

The result can be seen in the right panel of Figure 4–33. Remember, we are using the Polyline here to simplify the exercise. In your own models, you can choose not to draw a Polyline and sculpt the In-Place Edit Profile directly using the same methods and coordinate entry instead. Either approach is valid. In the next sequence, we will add our final Sweep using the Polyline directly.

Convert a Polyline to a Sweep

Another Polyline on the A-Wall-Adtv Layer has been provided in the Second Floor Construct as well. This one will be used to sculpt the Frieze Molding. Let's add it directly as a Sweep now.

28. On the Project Navigator Palette, on the Constructs tab, double-click to open the file named *Second Floor* within the *Constructs* folder.

 NOTE If you left the *Second Floor* Construct open above, then this action will simply make that file active.

29. **Thaw** the Layer named **A-Wall-Adtv**.

30. Select all three façade Walls (as we did above), and then Zoom in on the magenta Polyline at the Frieze Molding Wall Component.

31. With the three Walls selected, right-click and choose **Sweeps > Add**.

32. In the Add Wall Sweep dialog, from the Wall Component list, choose **Frieze Molding**. Leave Profile Definition set to **Start from scratch**.

33. For the New Profile Name, type **Frieze Molding Profile**, making sure that there is a checkmark in the **Miter Selected Walls** and the **Apply Roof/ Floor Lines to Sweeps** checkboxes, and then click **OK**.

34. At the "Select a location on wall for editing" prompt, click a point near the Polyline.

TIP Don't click directly on the polyline, but close to it. You have to click on the Wall for it to accept your pick.

Be sure that the cyan-colored In-Place Profile is selected (you should see its grips). If it isn't, click on it to select it before continuing.

35. Right-click and choose **Replace Ring**.

36. At the "Select a closed polyline, spline, ellipse, or circle" prompt, click on the magenta Polyline.

37. At the "Erase layout geometry" prompt, right-click and choose **Yes**.

38. Click the **Save All Changes** icon on the In-Place Edit toolbar to complete the Sweep (see Figure 4–34).

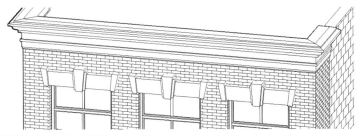

Figure 4–34 *The completed Frieze molding*

39. **Save** the file.

It might be nice to see how the overall model is shaping up at this point. Be sure that all open files are saved. (You can also close all open files and save them if you wish.)

40. On the Project Navigator Palette, on the Views tab, double-click to open the file named *Composite Model* within the *Views* folder.

41. From the View > Shade menu, choose either **Hidden** or **Gouraud Shaded**. (You can also use the icons on the Shading Toolbar.)

We have added some very nice touches to the façade with these Sweeps. However, Sweeps don't always work to give us the result that we need. For instance, similar treatment is required at Molding 1. However, since the first floor doors cut through Molding 1, the Molding profile would be cut square at these openings. However, a nicer treatment would be to allow the moldings to wrap back into the openings. This is not possible with Sweeps. To create this effect, we need to apply the molding to the façade as a separate object. We will do this in the next exercise.

42. **Save** and **Close** the *Composite Model* Construct file.

TRIM AND MOLDINGS USING STRUCTURAL MEMBERS

As we said above, to really control the exact position and path of your moldings, you need to apply them as a separate object. There are several techniques that could be used to achieve the effect we require: Mass Elements, Solids, or a Custom Wall Style. We are going to explore the use of Structural Members for this purpose.

Structural Members (Extruding Along a Path)

For this simple exercise, we will use a predefined Structural Member Style provided with this dataset and apply it to Polylines that match the path of our moldings. First we must draw the Polylines.

1. On the Project Navigator Palette, on the Constructs tab, double-click to open the file named *First Floor* within the *Constructs* folder.

 NOTE If you left the *First Floor* Construct open above, then this action will simply make that file active.

This part will be a little tricky because we need to trace some of the existing geometry, and it may be difficult to snap to just the right point. The required Polylines have been provided in a separate file if you get stuck. Try to create them yourself first.

2. From the View menu, choose **Shade > 2D Wireframe** (or pick the 2D Wireframe icon on the Shading Toolbar).

3. Zoom in on the Molding 1 Wall component (at the arch spring lines) on the right side of the façade.

4. Toggle off the Surface Hatch display.

 TIP Use the Surface Hatch Toggle icon on the Drawing Status bar to quickly toggle the surface hatch on or off.

5. Trace the top outside edge of the Molding 1 component, starting from the façade Wall in the alley and wrapping around the front, and turning back in 3 1/4″ to the face of the Window (see Figure 4–35).

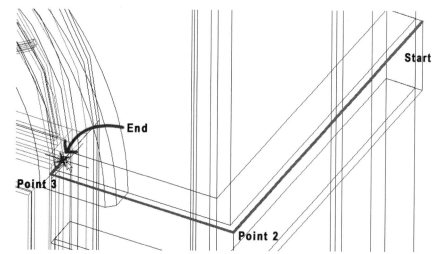

Figure 4–35 *Trace the top edge of the Modling 1 component of the façade Wall*

6. Repeat this process for each of the other sections of molding, moving from right to left across the façade Wall (see Figure 4–36).

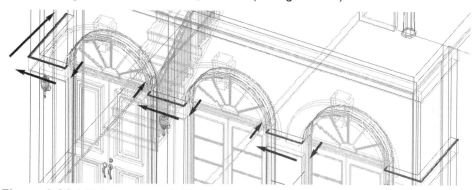

Figure 4–36 *Add the remaining Polylines*

If you have trouble getting these Polylines drawn correctly, you can insert the ones from the provided file instead. To do this, locate the file named *Molding 1 Paths* on the *Elements* folder, right-click it, and choose Insert as Block. In the Insert dialog, uncheck Specify on Screen for Insertion Point, Scale, and Rotation, and place a checkmark in the Explode checkbox. Click OK to import the Polylines into the file.

7. On the Project Navigator Palette, on the Constructs tab, double-click to open the file named *Moldings* within the *Elements* folder.

8. On the Tool Palettes, click the MasterVIZr tab to make this Tool Palette active.

The MasterVIZr Tool Palette was created above in the "Create the Second Floor Balcony Railing" heading.

9. On the Format menu, choose **Style Manager**, expand Architectural Objects, and highlight the Structural Members item.

10. On the right side of the Style Manager, drag the Style named **Molding 1** from the Style Manager to the MasterVIZr Palette. When the cursor changes to a plus (+) sign, release the mouse.

11. From the small menu that appears on the Tool Palette, choose **Beam**. Click **OK** to dismiss the Style Manager.

12. On the Project Navigator Palette, on the Constructs tab, double-click to open the file named *First Floor* within the *Constructs* folder.

 NOTE If you left the *First Floor* Construct open above, then this action will simply make that file active.

13. Select all four Polylines drawn (or imported) above.

14. Right-click the new Molding 1 Tool on the MasterVIZr Tool Palette and choose **Apply Tool Properties to > Linework**.

15. At the "Erase layout geometry" prompt, right-click and choose **Yes** (see Figure 4–37).

Figure 4–37 *The Polylines converted to Structural Members make very fine Moldings*

You should now have four moldings in place of the Polylines. The result here is very much like extruding a shape along a path. You could, of course, use Auto-CAD Solids and achieve virtually the same effect. But as was mentioned above, AutoCAD Solids do not use ADT Materials or the Display System. If any of the moldings appear "flipped," then this indicates that the Polylines where drawn left to right, rather than right to left. This is not a hard rule for Structural Members, but rather an aspect of the way that this particular Style was designed. If this occurred, you must undo it. There is unfortunately no way to "flip" or "reverse" them later. This is a limitation of this technique. To redo, you must draw the Polyline over again starting from the opposite end.

CREATE A CUSTOM STRUCTURAL MEMBER STYLE

In this sequence, we will add a dentil molding to the cornice on the second floor. Although, with some effort, we could model such a thing with Mass Elements, Structural Members are actually (as we have already seen) incredibly useful for many "non-structural" purposes, including dentil moldings! Unlike the previous exercise, in this one we will build the Structural Member Style from scratch.

CREATE A CUSTOM MEMBER SHAPE

To create a Structural Member Style, we first need a Custom Member Shape Definition. Unlike other ADT objects, Structural Members do not use AEC Profiles for their shape. However, the concept is the same. One or more Custom Member Shapes are extruded along the path of the Structural Member. Structural Members can also be used as Columns, Beams, or Braces. It makes little difference which is chosen when we are using them for moldings, but typically you will want to use a Beam as a molding. The first step is to draw the geometry that will be used for the shape(s), and then create a custom Member Shape from the command on the Format menu.

    1. On the Project Navigator Palette, on the Constructs tab, double-click to open the file named *Second Floor* within the *Constructs* folder.

 NOTE If you left the *Second Floor* Construct open above, then this action will simply make that file active.

Another very interesting aspect of Structural Members is their ability to contain three levels of detail per shape. In this way, the same Member can display differently in Low, Medium, and High Detail Display Configurations. For the Style that we are building here, we will use the same shape (a simple square) for both the Plan and Plan High Detail Displays. We will not configure the Plan Low Detail Display at this time, as it is not used when viewing the Model in 3D. Later, we will add the dentil teeth to make the Style more ornate in the High Detail Display.

    2. Draw a simple 1″ by 1″ square Polyline anywhere in the drawing, and then Zoom in on it. (You can change to Top view first if you like.)

    3. On the Format menu, choose **Structural Members > Member Shapes**.

4. Right-click and choose **New** (or type **N**and then press ENTER) to create a new shape.

5. Name the new style **Dentil** and then press ENTER.

6. Right-click on the new style and choose **Edit**.

7. On the General tab, type **Dentil Molding** for the Description.

8. On the Design Rules tab, on the right, select Medium Detail and then click the "Specify Rings for Medium Detail Shape" icon (see Figure 4–38).

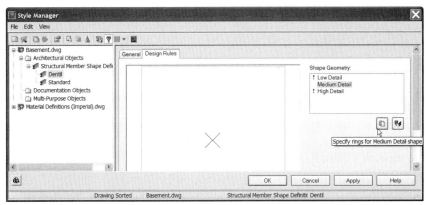

Figure 4–38 *Create a Custom Member Shape and Define the Rings for Medium Detail*

9. At the "Select a closed polyline, spline, ellipse, or circle for an outer ring" prompt, click the square drawn in step 2.

10. At the "Insertion Point" prompt, click the upper-left endpoint of the square (see Figure 4–39).

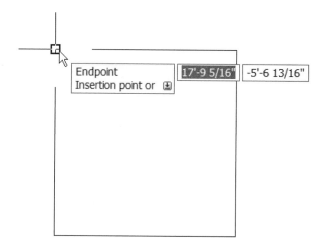

Figure 4–39 *Complete the Plan graphics by selecting the upper-left corner as the insertion point*

The Style Manager will return. If you made any mistakes while configuring the Medium Detail graphics, you can repeat the process to correct it. You can also choose Plan Low Detail, or Plan High Detail next to configure them. As we stated above, we do not need to worry about Plan Low Detail for this Structural Member Shape; however, we do want to configure Plan High Detail.

11. On the Design Rules tab, on the right, select **Plan High Detail** to configure the graphics for the Plan Display Rep.

12. Repeat the steps above using the same Polyline and Insertion Point as Medium Detail.

Normally, we would have a different shape for Plan High Detail. But for this exercise, what will make Plan and Plan High Detail different is the accoutrement added to High Detail, not the section profile shape.

13. Click OK to close the Style Manager.

14. Erase the square Polyline and then **Save** the drawing.

Create the Dentil "Teeth" Display Block

The "teeth" that will display along the Model High Detail Display will be a simple Custom Display Block. Structural Members can use Display Blocks just as Doors, Windows, and Railings did above.

15. Be sure the Design Tool Palette Group is active (right-click the Tool Palettes title bar and choose Design if it is not), and then click the **Massing** tab.

16. Click the **Box** Tool. On the Properties Palette, from the Specify onscreen list, choose **No**.

17. Set the Width to **2″**, the Depth to **3 1/2″**, and the Height also to **3 1/2″** (see Figure 4–40).

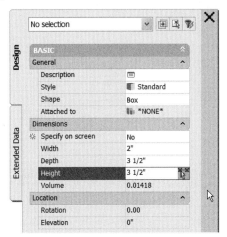

Figure 4–40 *Configure the sizes of the Box on the Properties Palette, rather than Specify onscreen*

18. At the "Insertion point" prompt, click a point anywhere onscreen, and then press ENTER to accept the default Rotation of 0°. Press ENTER again to complete the command.

19. From the View menu, choose **3D Views > SE Isometric** (or click the icon on the Views toolbar).

20. Zoom in on the Box. Select the Box, and on the Properties Palette, change the Style to **Arch**.

21. With the Box still selected, right-click and choose **Copy Mass Element Style and Assign**.

22. On the General tab, name the new Style **Dentil Molding**, and then click **OK**.

23. Create a Block named "Dentil Tooth" from the Box, with the Midpoint of the top short edge as the Base Point (see Figure 4–41).

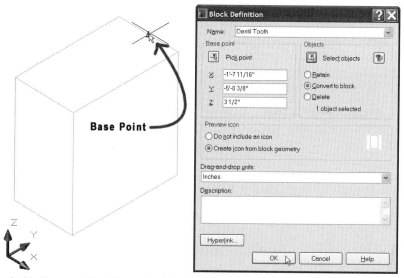

Figure 4–41 *Create a Block from the Mass Element Box*

24. Select the Block, right-click, and choose **Set Nested Objects to ByBlock**.

ByBlock is a special setting that makes the objects within a Block or ADT Object use the display properties of the whole.

25. Erase the Block onscreen, and return to the previous Zoom of the façade. (SE Isometric rotated a little off-center to a more "head-on" view.)

26. **Save** the File.

Create a Structural Member Style

We will now create a new Structural Member Style that uses the new shape and the Custom Display Block.

27. On the Design Tool Palette, right-click the Structural Beam Tool and choose **Structural Member Styles**.

28. Create a **New** Style named **Dentil Molding** and then **Edit** it.

29. Click the Design Rules tab, and rename the "unnamed" component to Section, and then choose **Dentil** from the Start Shape Name list (see Figure 4–42).

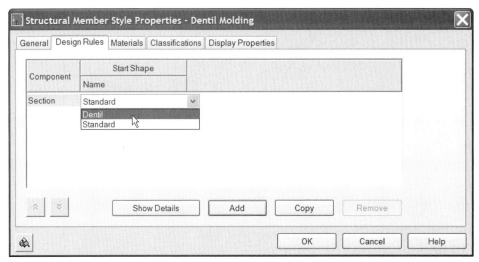

Figure 4–42 *Assign the previously defined Custom Member Shape to the Section of the new Structural Member Style*

30. Click the Materials tab. Choose **Finishes.Plaster and Gypsum Board.Plaster.Stucco.Fine.Brown** as the Material for the Section component.

31. Click the Display Properties tab, and place a checkmark in the Style Override box next to **Model High Detail**.

32. In the Display Properties dialog, click the Other tab, and then click the **Add** button in the **Custom Block Display** area.

Custom Blocks are used in many Styles to represent items that cannot be represented parametrically. We used them in several examples above. In this exercise, we will use the display Block Dentil Tooth that we just defined to represent the "teeth" on the molding. Since we are applying this edit to Model High Detail, these teeth will only appear when the High Detail Display Configuration is active.

33. Click the **Select Block** button, choose **Dentil Tooth** from the Block list, and then click **OK**.

34. In the Position Along (X) area, choose **Midpoint of Curve**.

This will center the teeth, starting from the middle of the total molding length.

35. In the Insertion Offset area, type **1″** in the **y** field.

This shifts the Block relative to the Structural Member. The amount of the offset is determined based on the size of the Custom Member Shape and the Base Point of the Block.

36. Place a checkmark in the **Repeat Block Display** box.

37. In the **Space Between** field, type **4″**.

38. Place a checkmark in the **Fit** box, and choose the **Scale to Fit** option (see Figure 4–43).

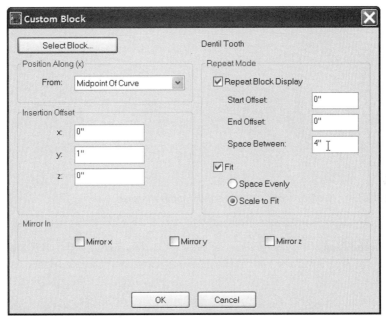

Figure 4–43 *Configure the repeating Block display to show "teeth" along the molding*

39. Click **OK** four times to return to the drawing.

40. **Save** the drawing.

Add the Dentil Molding Style to the Façade

All that remains is to use the Style on the façade. To do this, we will again convert a Polyline to a Beam. The Polyline has been provided to expedite the process.

41. From the Views menu, choose **3D Views > SE Isometric**. (Or use the SE Isometric view icon on the Views toolbar.)

42. From the View menu, choose 3D Orbit. Place your cursor over the small circular handle at the bottom of the green circle and drag upwards until you can see underneath the Cornice Molding (see Figure 4–44).

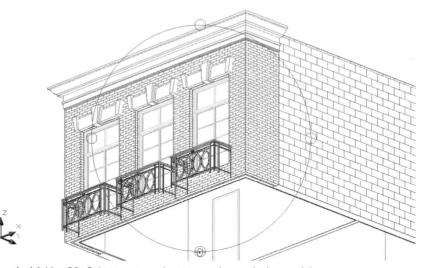

Figure 4–44 *Use 3D Orbit to rotate the view underneath the model*

43. Zoom in on the Cornice Molding.

44. On the Project Navigator Palette, on the Constructs tab, right-click the file named *Freeze Molding Path* in the *Elements* folder, and choose **Insert as Block**.

45. In the Insert dialog, uncheck all "Specify on Screen" checkboxes and place a checkmark in the **Explode** checkbox, and then choose OK.

You should see a magenta Polyline appear along the Cornice Molding.

46. Select the magenta Polyline just inserted.

47. On the Design Tool Palette, right-click the Structural Beam Tool and choose **Apply Tool Properties to > Linework**.

48. At the "Erase layout geometry" prompt, right-click and choose **Yes**.

A large boxy Structural Member will appear and remain highlighted. Do not deselect it.

49. With the new Structural Member still selected, change the Style on the Properties Palette to **Dentil Molding**.

50. Change the Justify to **Top Right**.

51. On the Drawing Status Bar, choose **High Detail** from the Display Configuration pop-up menu (see Figure 4–45).

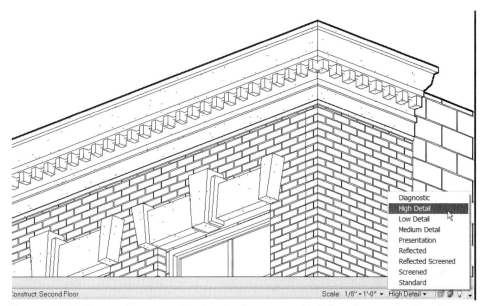

Figure 4–45 *Change the current Display Configuration to High Detail to see the Dentil Molding*

Notice that in High Detail, the teeth appear. The hatching on the Walls also gets more detailed. When you build a Structural Member this way, you now have the option to easily eliminate the highly detailed dentil molding when you are viewing the model from afar. This will help to speed up rendering times in VIZ Render by not forcing the rendering engine to calculate unnecessary details.

52. **Save** the file.

WALL OPENING ENDCAPS

In Chapter 3, we removed all Endcaps from the various Wall Styles in use in this project and replaced them with the Standard Endcap Style. Standard is a simple straight line that terminates each Wall component perpendicular to the face of the Wall. By customizing Endcaps, we can make the ends of the Walls terminate any way we wish. In this sequence we will put the finishing touches on the Windows of the second floor. By using a custom Endcap condition, we can control the way the brick wraps into the Windows that we previously set back within the Wall. On the first floor, we will use an alternative technique to achieve a similar goal.

Using Auto-Calculate Endcap

Endcaps can be tricky to master. Fortunately, ADT can calculate the Endcap settings automatically. This makes creating them much easier. To begin, we need some open Polylines; one for each component in the Wall Style that must be capped.

1. On the Project Navigator Palette, on the Constructs tab, double-click to open the file named *Second Floor* within the *Constructs* folder.

NOTE If you left the *Second Floor* Construct open above, then this action will simply make that file active.

2. From the View menu, choose **3D Views > Top** (or click the Top view icon on the Views toolbar). Also make sure that shading is turned off (choose **2D Wireframe**).

3. On the Design Tool Palette, choose the Wall tool. On the Properties Palette, change the Style to **Townhouse Second Floor Facade Walls** and set the Base Height to **12'-6"**.

4. Draw a small left-to-right horizontal Wall segment away from the main model, about 4'-0" long (see Figure 4–46).

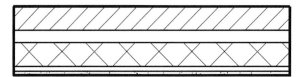

Figure 4–46 *Draw a short horizontal Wall segment*

5. Draw a Polyline starting at the point indicated in the leftmost panel of Figure 4–47, and following the dimensions shown (see Figure 4–47).

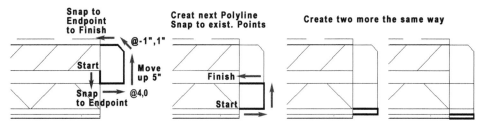

Figure 4–47 *Draw a Polyline for the Brick*

6. Create three more Polylines following the same process as shown in the three remaining panels to the right in Figure 4–47.

The dashed line that appears above the top edge of the Wall is the Cornice component above our heads. Even though we do cut through it in plan, we still need to account for it in the Endcap Style. Therefore, we must draw one more Polyline.

7. Draw another Polyline, as shown in Figure 4–48.

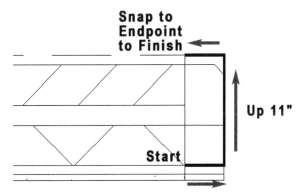

Figure 4–48 *Draw the Polyline for the Cornice Molding Component above*

8. Move all the Polylines to the left **4"**. (They should now overlap the end of the Wall.)

9. Select the Wall, right-click, and choose **Endcaps > Calculate Automatically**.

10. At the "Select polylines" prompt, select all five Polylines just drawn and then press ENTER.

11. At the "Erase selected polyline(s)" prompt, right-click and choose **No**.

We will erase them later, but we will need some of these Polylines again.

12. At the "Apply the new wall endcap style to this end as" prompt, right-click and choose **Override**.

13. In the small dialog that appears, type **Turned In Brick** for the name, and then click **OK** (see Figure 4–49).

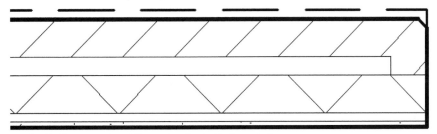

Figure 4–49 *The completed Turned in Brick Endcap Style*

This is the Endcap that we will ultimately apply to the jambs of the second floor Windows. We need to create one more for the sills. We are going to cheat a little here. We are going to allow the CMU component to "flow" out and become the sill. This will work OK for us since the CMU component can be assigned a different render Material without adversely affecting its Plan display. We will see this in later chapters.

14. Mirror the Polylines from the right side of the Wall to the left side. Use the Midpoint of the Wall as the mirror point so that the Polylines end up in the precise location on the other end.

15. Delete the chamfered brick Polyline and the "U"-shaped Polyline for the CMU from the left side.

16. Using the same process as the other Polylines, draw a new Polyline on the left, as shown in Figure 4–50.

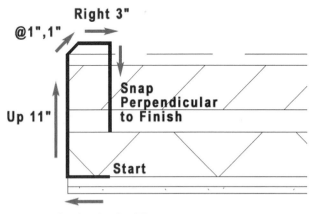

Figure 4–50 *Draw a new Polyline for the Sill*

17. Repeat steps 9 through 12.

18. Name the new Endcap Style **Window Sills**, and then click **OK** (see Figure 4–51).

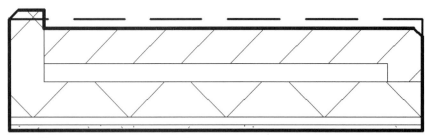

Figure 4–51 *The resultant Window Sill Endcap*

19. Delete the 4'-long Wall and all of the polylines.

There are two types of Endcaps in ADT. Wall Endcaps, which are used at the ends of freestanding Walls, and Wall Opening Endcap Styles, which are used at the holes made by Doors, Windows, Assemblies, and Openings. Wall Opening Endcap Styles have a Wall Endcap Style assignment at each Jamb, the Sill, and the Head. This is why we had to first create our Wall Endcap Styles. As you saw, we can use those Endcap Style on the Wall ends, but really they were designed for the

Windows. So let's make a Wall Opening Endcap Style now that references both of the Endcap Styles just created.

20. From the Format menu, choose **Style Manager**, expand Architectural Objects, and then select **Wall Opening Endcap Styles**.

21. Right-click on the right side, and choose **New**. Name the new Style **Second Floor Windows**.

22. Double-click the new style to edit it.

23. Click the Design Rules tab.

24. Choose **Turned In Brick** for Jamb Start and Jamb End. Choose **Window Sills** for Sill.

25. Click **OK** twice to return to the drawing.

26. Select the main horizontal façade Wall, right-click, and choose **Edit Wall Style**.

27. Click on the Endcaps/Opening Endcaps tab, and choose **Second Floor Windows** for the Door/Window Assembly assignment in the Opening Endcaps area.

28. Click **OK** to return to the drawing (see Figure 4–52).

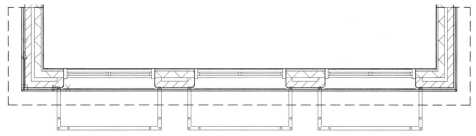

Figure 4–52 *Assign the new Opening Endcap Style to the Door/Window Assembly slot of the Townhouse Second Floor Facade Walls Style*

29. From the Views menu, choose **3D Views > SE Isometric**. (Or use the SE Isometric view icon on the Views toolbar.)

Zoom in on the Windows. If you still have High Detail set as your active Display Configuration, then the Opening Endcaps should display. However, if you change back to Medium Detail, notice that the Endcaps do not appear in 3D Model (see Figure 4–53). If you wish to have them display in both Medium and High Detail, select the Wall, right-click, and choose Edit Object Display. Highlight Model and then click the Edit Display Properties icon. Click the Other tab, and place a checkmark in the Display Opening Endcaps checkbox. This is not really necessary for this model, since we will be using the High Detail Display Configuration in later chapters to render. If you make changes to the Display Configuration and the change does not seem to occur, use the Regenerate Model icon on the Standard toolbar to update the screen display.

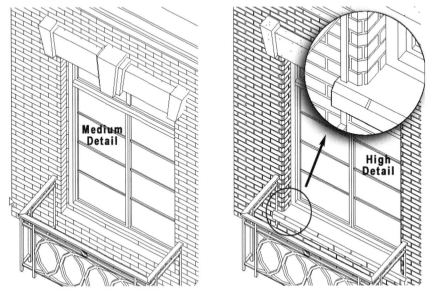

Figure 4–53 *Opening Endcaps Display by default only in High Detail*

    30. **Save** the file.

CHAMFERED BRICK ON FIRST FLOOR

It is desirable to have the same chamfered brick effect on the first floor. Unfortunately, however, the complexity of the Wall Style used on the First Floor file and the fact that we have half-round openings will prevent us from achieving successful results with Opening Endcap Styles. Wall Opening Endcap Styles typically work well with rectangular openings, but tend to fail with other shapes. Instead, we will use Body Modifiers to "carve" away the brick at the lower rectilinear portion of the French Doors. To expedite the process, Extruded Mass Elements have been provided already in a separate Element file. These Mass Elements were created from triangular-shaped Polylines, and then extruded with the Convert to > Mass Element right-click option. Feel free to build them yourself if you prefer.

Adding Body Modifiers

    1. On the Project Navigator Palette, on the Constructs tab, double-click to open the file named *First Floor* within the *Constructs* folder.

NOTE If you left the *First Floor* Construct open above, then this action will simply make that file active.

2. From the View menu, choose **3D Views > Top** (or click the Top view icon on the Views toolbar). Also make sure that shading is turned off (choose **2D Wireframe**).

3. On the Project Navigator Palette, on the Constructs tab, right-click the file named *Brick Chamfers* in the *Elements* folder, and choose **Insert as Block**.

4. In the Insert dialog, uncheck all "Specify on Screen" checkboxes, and place a checkmark in the **Explode** checkbox.

You should see six small triangular-shaped Mass Elements appear, two by each Door.

5. Select the main horizontal façade Wall, right-click, and choose **Body Modifiers > Add**.

6. At the "Select objects to apply as body modifiers" prompt, carefully select all six triangular-shaped Mass Elements, and then press ENTER. (Be sure not to select any objects other than the six Mass Elements.)

7. In the Add Body Modifier dialog, choose **Brick 1** from the Wall Component list.

8. Choose **Subtractive** from the Operation list.

9. Type "**Chamfer Brick Corners**" in the Description field.

10. Place and checkmark in the **Erase Selected Object(s)** checkbox, and then click OK (see Figure 4–54).

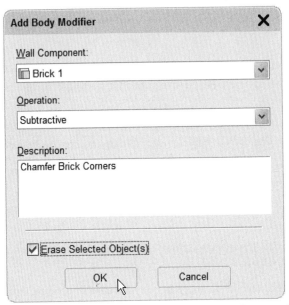

Figure 4–54 *Apply the six triangular-shaped Mass Elements as subtractive Body Modifiers*

11. From the View menu, choose **3D Views > SE Isometric** (or click the SE Isometric view icon on the Views toolbar).

12. Save the file.

Notice that the effect is very similar to the chamfered bricks on the second floor. It is important to note that with this technique, the chamfered effect cannot be toggled on and off when changing Display Configurations as it can with Endcaps. Body Modifiers change the actual geometry of the Wall, regardless of the active Display Config.

View the Final Model

Our Townhouse model is now complete. We will load our Composite Model one more time and have a look at the final results. Save and close all Constructs before continuing.

13. On the Project Navigator Palette, on the Views tab, double-click to open the file named *Composite Model* within the *Constructs* folder.

14. From the View menu, choose **3D Views > SE Isometric** (or click the SE Isometric view icon on the Views toolbar).

15. Use 3D Orbit to adjust the view.

16. On the Drawing Status Bar, choose **High Detail** from the Display Configuration pop-up menu.

17. From the View menu, choose Shade > **Gouraud Shaded** (or choose the Gouraud Shaded icon from the Shading toolbar). (See Figure 4–55 and Color Plate C–4b.)

Figure 4–55 *The complete Townhouse Composite Model View file*

18. **Save** the file.

CONCLUSION AND SUMMARY

In this chapter, we have completed our townhouse model. We explored several different modeling techniques within Autodesk Architectural Desktop. We built balcony Railing Styles, and applied Custom Display Blocks to them to make them more interesting and ornate. Using similar techniques, we spruced up the Window Styles created in Chapter 3 with Custom Display Blocks for Arches. We added Door Hardware and exterior lighting. We also looked at several different approaches to trim and moldings, including Sweeps and Structural Members. We saw how certain modeling techniques can be made to respond to the current Display Configuration. We completed our façade with the application of Endcaps and Body Modifiers.

As you can see, there is a very robust toolset available within ADT to create very rich and complete Models. In two of the chapters that follow, we will revisit this model again, and bring it to life with all the powerful tools available in VIZ Render. Looking back on what we have accomplished in this chapter, you learned how to:

Create Railing Styles.

Add Custom Display Blocks to railings.

Apply a Railing Style to a Polyline.

Build a Custom Display Block.

Create Revolution Mass Elements.

Apply Custom Display Blocks to windows and doors.

Edit Door Profiles In-Place.

Work with Sweeps.

Edit Sweep Profiles In-Place.

Convert a Polyline to a Sweep.

Convert a Polyline to a Structural Member.

Build a Custom Member Shape.

Build a Custom Structural Member Style.

Work with Wall Endcaps.

Create Endcaps automatically.

Create a Wall Opening Endcap Style.

Apply a Body Modifier to a wall.

CHAPTER 5

Exterior Daylight Rendering

INTRODUCTION

As we mentioned in the previous two chapters, before any rendering can take place, we must have a model from which to generate our presentations. The model that we will use in this chapter and in Chapter 7 is the model that we constructed in Chapters 3 and 4—The Townhouse. If you did not complete those chapters, and wish to understand the various modeling techniques that were used to construct the dataset used in this chapter, please return to Chapters 3 and 4 now and complete them first. However, you are not required to build the model yourself, or even complete Chapters 3 and 4 at all in order to complete this chapter. Feel free to start right here if you are anxious to get right to rendering. In this chapter, we will explore the creation of an exterior daylight rendering. We will add a Daylight system and render the Townhouse façade using a variety of camera angles.

OBJECTIVES

The primary objective of this chapter is to produce a simple exterior daylight rendering from a variety of vantage points. Supplemental to this goal will be the understanding of cameras and the third-party Bionatics plug-in to add trees to our scene. In this chapter, we will explore:

- Setting up a linked model for rendering
- Creating and manipulating Cameras in VIZr
- Adding Entourage to the Scene
- Creating an Exterior Daylight Rendering

THE TOWNHOUSE DATASET

The project for this chapter is a two-story townhouse. The model is provided as a series of Constructs in the ADT Project Navigator. All of the Constructs have been provided, including the façade of the townhouse which was modeled in Chapters 3 and 4. If you did not complete those chapters, you can visit them now or later to gain a complete understanding of how the model was constructed.

In this chapter, we will use a basic color environment background, explore various Cameras including a camera correction modifier, and create a simple Daylight System (to represent the Sun). If you completed the previous chapters, you'll recall that there are many details in this façade that will cast some very dynamic shadows. We will keep this in mind as we set up our Daylight System. Something else to keep in mind is that this site is a bit tight. Because of this, we will have to position the camera fairly close to the building. These dynamics present an opportunity to look at creative camera positioning and lighting scenarios.

Start Autodesk Architectural Desktop and Load the Townhouse Project

1. If you have not already done so, install the dataset files located on the Mastering VIZ Render CD ROM.

   Refer to "Files Included on the CD ROM" in the Preface for information on installing the sample files included on the CD.

2. Launch Autodesk Architectural Desktop from the icon on your desktop.

   You can also launch ADT by choosing the appropriate item from your Windows Start menu in the Autodesk group. In Windows XP, look under "All Programs"; in Windows 2000, look under "Programs." Be sure to load Architectural Desktop, and not VIZ Render, for this exercise.

3. From the File menu, choose **Project Browser**.

4. Click to open the folder list and choose *My Computer* and then your C: drive.

5. Browse to the *C:\MasterVIZr* folder.

6. Double-click **Townhouse05** to make this project current (you can also right-click on it and choose **Set Current Project**). Then click Close in the Project Browser.

 IMPORTANT *If a message appears asking you to re-path the project, click Yes. Refer to the "Re-Pathing Projects" heading in the Preface for more information.*

CREATE A VIEW DRAWING TO BE USED FOR RENDERING

As we saw in the previous two chapters, a View file in Project Navigator gathers one or more of the Constructs of the project and assembles them at the correct physical locations and heights. Views are often used to present selected portions of the model with the required annotations and embellishments for construction documents. However, we can also use Views to compile composite models of our project for rendering. For example, if you were only doing an exterior rendering and your project was set up with shell and interior constructs, you would create a

View drawing of the site and shell Constructs only. For this exercise we will create a complete composite model for rendering.

> When you loaded the *Townhouse05* project in the Project Browser above, the Project Navigator Palette should have opened automatically. If it did not, please load **Project Navigator** from the Window menu now.

7. On the Project Navigator Palette, click on the Views tab.

8. On the Views folder, right-click and choose **New View Dwg > General**.

9. For the Name, type **Rendering – Front Perspective**, for the Description, type **Composite Model for Rendering,** and click OK to exit the Description dialog. Then click Next.

 NOTE Although our project does already contain a Composite Model View file, it is often good practice to create a separate View file for each distinct project requirement or drawing type. In this way, we can freely modify our Rendering View file as necessary for rendering, without adversely affecting the existing View, whose purpose might be for CDs or project coordination, which often have needs quite different from rendering.

10. On the Context page, right-click on the right side, choose **Select All**, and then click Next.

> This will select all Levels to be included in the composite model.

On the Content page, notice that all Constructs are selected for inclusion in the model. This occurs since we selected all Levels on the previous screen. ADT will search your model based on your choice of Context and offer to include all Constructs it finds which meet that criteria. You can optionally remove the checkmark from any Construct to manually remove it from the View.

11. Click Finish to complete the creation of the View.

12. In the Project Navigator, double-click on the newly created View, **Rendering – Front Perspective**, to open it.

13. On the Views toolbar, click the SW Isometric icon (or choose **3D Views > SW Isometric** from the View menu).

14. On the Shading toolbar, click the Gouraud Shaded icon (or choose **Shade > Gouraud Shaded** from the View menu).

15. From the Display Configuration pop-up menu on the drawing status bar, choose **High Detail** (see Figure 5–1).

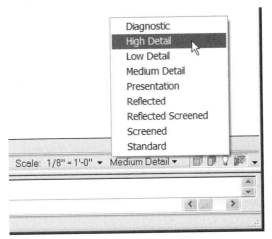

Figure 5–1 *Change the current Display Configuration*

We are now ready to link our file over to VIZ Render and begin creating an exterior daylight rendering.

16. Zoom in on the front façade and then Save the file.

SET UP THE RENDERING

As we have seen in previous chapters, models are built in ADT and then linked to VIZ Render to create renderings and animations. Be sure to save your ADT model and then initiate the linking process from directly within ADT. While a file link can be reloaded from VIZ Render, it cannot be created in VIZr.

Link to VIZ Render and Configure the Viewports

The "Link to VIZ Render" command is on the Open Drawing Menu (accessed from the small circular icon on the Drawing Status Bar, see Figure 1–2 in Chapter 1.)

1. On the Drawing Status bar, click the "Open Drawing Menu" icon (a small triangle within a circle on the left), and choose **Link to VIZ Render** from the menu that appears (illustrated in Figure 1–2 in Chapter 1).

The VIZ Render application will launch, and after a short delay, the model will appear within the VIZ Render window (see Figure 5–2).

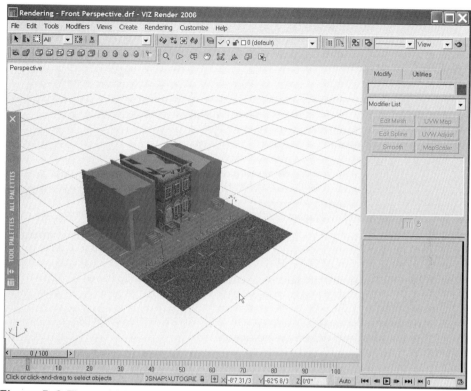

Figure 5–2 *The model loads in VIZ Render*

2. On the Viewport Navigation Tools toolbar, click the Minimize Viewport icon (illustrated in Figure 2–6 in Chapter 2).

   You may also press ALT + W.

3. On the same toolbar, click the Zoom Extents All icon (illustrated in Figure 2–10 in Chapter 2).

Add a Camera

We will render three different camera views, each using a different technique to produce a final rendering.

4. Right-click in the Top viewport to make it active.

5. From the Create menu, choose **Cameras > Target Camera**.

   For the first camera angle, we'll try a shot from the left side.

6. Pick a point near the leftmost lamppost in the street, drag toward the townhouse, and release.

7. Watching your progress in the Front or Left viewport, drag up to about eye-level, and then click to complete placement of the Camera (see Figure 5–3).

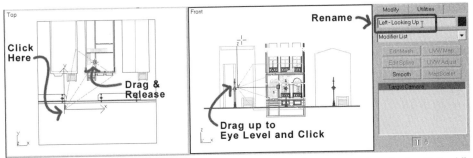

Figure 5–3 *Add a Camera by clicking its location and then dragging to where it should "look"*

At the moment we are not too concerned with the exact placement of the Camera as we will be adjusting the angle in upcoming steps.

8. With the Camera01 still selected, rename it (on the Modify Panel) "**Left – Looking Up**" (see Figure 5–3).

9. Activate the Perspective viewport and set it to the "Left – Looking Up" Camera view (press **C** on the keyboard).

The camera view is a bit too close. At this point, we have four options. One, change the zoom factor (FOV) of the camera. Two, adjust the proportions of the rendering. Three, change the camera location. Or four, a combination of the other three. Let's go with option Four; however, let's look at each one.

Adjust the Field of View

The first two parameters, Lens and FOV, are linked together. The Lens parameter uses a designation in millimeters to emulate a zoom lens of an actual 35mm camera. In 35mm photography, a 50mm lens closely approximates the field of view of the unaided eye. Any focal length smaller than 50mm is considered a wide angle lens, while any focal length higher is considered a telephoto lens. A lens that can adjust from one focal length to another is called a "zoom" lens. This is the type of lens available to us in VIZr. Wide angle lenses take in more of the scene, or a view "wider" than is possible with the unaided eye. A telephoto lens offers a narrower view but is capable of bringing objects from a greater distance into view. The Field of View (FOV) setting is inversely proportional to the Lens setting. It measures the included angle of the chosen focal length. The wider the angle of view, the smaller the number of millimeters and the greater the included angle in the FOV. Conversely, the narrower the angle of view (telephoto) the larger the millimeter designation but the smaller the FOV angle. To make adjustments, use the spinners next to Lens and FOV, or click one of the Stock Lenses buttons.

Be sure that the Camera is still selected. If it is not, simply click on it in one of the orthographic views, or press the **H** key on the keyboard to select it by name in the Select Objects dialog.

10. In the Modify Panel, change the FOV settings by adjusting the spinners next to Lens or FOV.

Notice that both values change as adjustments are made to either spinner. Also notice the view in the Camera viewport adjust in real-time.

11. Click any of the buttons in the Stock Lenses section.

12. After trying various settings, click the 24mm Stock Lens button (see Figure 5–4).

Figure 5–4 *Adjust the Lens of the Camera by choosing a Stock Lens*

You should now have a view that shows most of the façade as well as the surrounding buildings and street.

Adjust Render Output Settings

Now that we have seen the effect of the Camera Lens and FOV settings, let's try changing the proportions of the output rendering.

Be sure that the "Left – Looking Up" viewport is active. If it is not, simply right-click in it before proceeding.

13. From the Rendering toolbar, click the Render Scene icon (see Figure 5–5).

Figure 5–5 *Click the Render Scene icon on the Rendering toolbar*

14. On the Common tab in the Common Parameters rollout, in the Output Size area, change the Width to **600** and the Height to **800**.

   Notice that there is no change in the "Left – Looking Up" camera viewport yet.

15. Right-click on the "Left – Looking Up" viewport label, and choose **Show Safe Frame** (see Figure 5–6).

 NOTE Make sure you right-click on the viewport label, not the viewport itself; if you right-click the viewport, a different set of menu commands will appear.

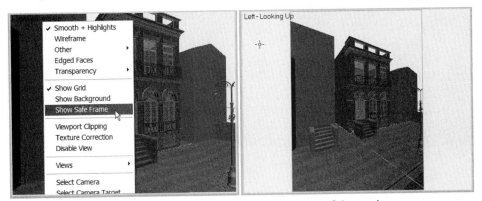

Figure 5–6 *Turn on "Show Safe Frame" to see the proportions of the render output*

   We are close to the viewpoint desired. We can make further adjustments to the position of the camera to get it just right.

16. From the Viewport Navigation Tools toolbar, press and hold down on the Orbit Camera icon (this will reveal a flyout toolbar). Choose **Pan Camera**.

17. In the camera viewport, pan the viewport so that you can see the top of the townhouse and the lamppost on the right.

   Right-click to cancel the Pan Camera action when you are satisfied (see Figure 5–7).

NOTE The coordinate location of our camera is X = –12', Y = –30', Z=4', and the Target is X=13', Y=4', Z=10'. At the end of our camera modifications we will provide the locations so that your renderings will look like ours if you wish. To do this, right-click with the Left – Looking Up camera selected and select Move. Right-click again and choose the Transform Type-In icon next to Move on the Edit/Transform Quad menu. Input the coordinates given here. Right-click again, and choose Select Camera Target from the Tools I Quad menu, and then input the coordinates for the Target in the Transform Type-In. For more information on these Move techniques, refer to the "Transforms" topic of Chapter 2. Feel free to adjust our coordinates to your satisfaction.

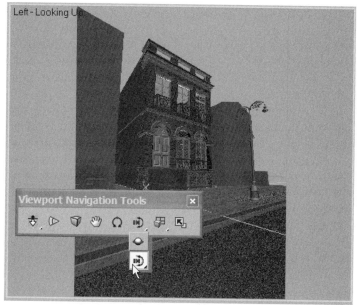

Figure 5–7 *Adjust the Camera view by using the Pan Camera tool*

Refer to the "Viewport Navigation" topic in Chapter 2 for more information on the Viewport Navigation tools.

18. From the File menu, choose **Save**.

ADDING DAYLIGHT

Since this is an exterior daytime scene, we will be using a Daylight System as the only light source. The VIZr Daylight System tool accurately simulates the light of the Sun from any place on Earth. Using the available parameters, we can adjust cloud cover, geographic location, and a host of other settings.

1. From the Create menu, choose **Daylight System**.

2. Click the **Yes** button in the Daylight Object Creation dialog to accept the recommended changes to your Environment settings (see Figure 5–8).

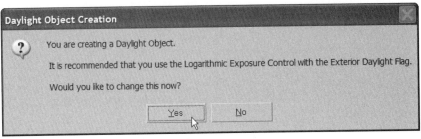

Daylight Object Creation

You are creating a Daylight Object.

It is recommended that you use the Logarithmic Exposure Control with the Exterior Daylight Flag.

Would you like to change this now?

Yes   No

Figure 5–8 *Accept the default Daylight Object parameters*

3. Using the steps outlined in Chapter 1, create a Daylight System in the Top viewport, with North pointing up.

NOTE Refer to the "Daylight" Heading in Chapter 1 for more information on placing a Daylight system.

Be sure that the Daylight object (Daylight01) is still selected.

4. On the Modify panel, in the IES Sky Parameters rollout, choose the **Partly Cloudy** radio button in the Coverage area.

When you are preparing a scene for rendering, you will often perform several test renderings as you fine-tune your settings. With this in mind, initially you will be concerned more with the speed of these test renderings rather than producing a final photorealistic product. The next several steps will showcase this process.

5. In the IES Sky Parameters rollout, within the Render area, set the Rays per Sample to **5**.

As you may recall from Chapter 1, this setting relates to how many rays are bounced off a surface. The more there are, the more accurate the results are however, more Rays can also increase processing time.

6. In the Control Parameters rollout, change the settings in the Time area by setting the Hours to **10**. Leave Minutes and Seconds set to **0**.

7. Leave the Month set to **6**, change the Day to **26**, and leave the year set to **2005**.

8. Place a checkmark in the Daylight Saving Time checkbox.

This Townhouse project is located in Chicago. Let's set that location for the Daylight System now.

9. In the Location area, click the Get Location button.

10. Choose **Chicago, IL**, for the location, and then click OK (see Figure 5–9).

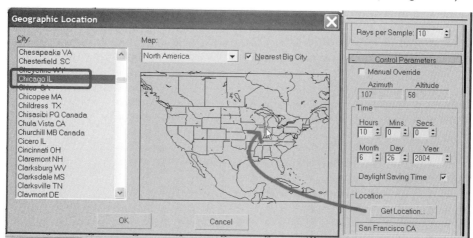

Figure 5–9 *Choose Chicago, IL, for the Location to adjust the Daylight System to this location*

If you have a really steady hand you can select a point on the map near where Chicago is (if you have "Nearest big City" checked) to set it as the location. If you miss the right spot, you can also choose it from the City list. If you are doing renderings outside of North America, you can choose a different map from the Map list first.

Since this scene has some reflective materials in it we should adjust one more setting in order to speed processing.

> If you closed the Render dialog above, from the Rendering toolbar, click the Render Scene icon (see Figure 5–5).

11. In the Render Options rollout, set the Ray Bounces to **2**, and then close the Render Scene dialog.

Ray Bounces is used when calculating reflections and refractions. The higher the number the more accurate glass will look at the cost of rendering time. Since we are in the "testing" phase, accurate-looking glass is not as important as being able to quickly set up a scene.

REFINING THE SCENE

At this point we have enough of the basic setup complete to begin creating renderings. We will address a few issues regarding the content in the rendering as we go along. The basic process is to create a rendering, evaluate the results, adjust the settings, and then render again. With each iteration, we will get closer to our desired final results. Let's get started and create the first rendering.

Rendering the Camera View

1. With the Camera viewport active, click the Quick Render icon on the Rendering toolbar (see Figure 5–10 and Render Stats 05.01).

Figure 5–10 *Perform your first quick render from the Camera Viewport*

Render Stats 05.01

Render Time: 34 seconds
Radiosity Processing Time: NA
Music Selection: "Symphony No. 9 (Scherzo)" Ludwig van Beethoven, composer.
Seattle Symphony. Gerard Schwarz, director – Beethoven's Symphony No. 9 (Scherzo)
Color Plate: NA
JPG File: Rendering 05.01.jpg

This is supposed to be a daytime rendering, so the first thing we need to adjust is the very dark background.

2. On the Rendering menu, choose **Environment** (or type **8** on the keyboard).

   This will open the Render Scene dialog and load the Environment tab.

3. Click on the Color swatch in the Background section.

4. Change the color to a light blue color and then click Close.

   Try Red: **206**, Green: **206**, Blue: **250** (see Figure 5–11).

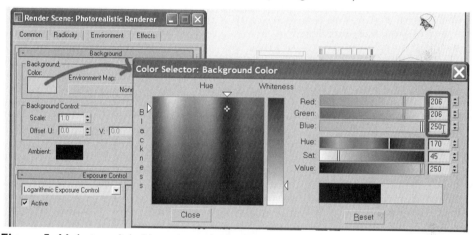

Figure 5–11 *Assign a light blue color to the background*

5. Close the Render Scene dialog.

Adjust the Material Properties

Now let's make some adjustments to the General Chip Material that is currently applied to the two neighboring houses.

6. On the Materials tool palettes, click the Scene – In Use tab.

7. Right-click on General Chip and choose **Properties**.

8. Expand the Special Effects rollout.

9. Click and drag the None button (next to Cutout) and drop it on top of the Bump Slot (see Figure 5–12).

NOTE You want to actually click and hold down on the "None" button and then drag it on top of the Map button next to the Bump slot. This process replaces the Map assigned to Bump with the one assigned to Cutout (which in this case is None).

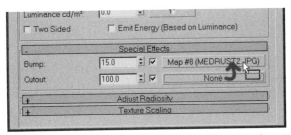

Luminance cd/m²   0.0
☐ Two Sided   ☐ Emit Energy (Based on Luminance)

	Special Effects		
Bump:	15.0	☐ ☑	Map #8 (MEDRUST2.JPG)
Cutout:	100.0	☐ ☑	None
+	Adjust Radiosity		
+	Texture Scaling		

Figure 5–12 *Dragging the item assigned to one slot to another to replace it with the item from the first*

This removes the unnatural texture from the neighboring houses. Let's do another rendering to see the effects of these two changes.

10. Close the Material Editor.

11. Click the Quick Render icon on the Rendering toolbar again (see Render Stats 05.02).

The blue sky helps considerably. Also, the neighboring buildings appear less distracting too, now that the texture has been removed (see Figure 5–13).

Figure 5–13 *Adding a blue sky and removing the texture on the neighbors helps*

Render Stats 05.02

Render Time: 36 seconds
Radiosity Processing Time: NA
Music Selection: "Chicago" Frank Sinatra, Nelson & Orchestra Riddle – The Capitol
Years: The Best of Frank Sinatra Disc 2 (1953)
Color Plate: NA
JPG File: Rendering 05.02.jpg

Clone the Camera

Perhaps we should try another camera angle. We can create a new Camera using the same procedure that we did above, or we can simply duplicate the Camera we have and modify its parameters.

12. Right-click in the Top viewport and then Maximize it (use the Maximize Viewport icon on the Viewport Navigation Tools toolbar, or press ALT + W).

13. Use the Zoom Region Mode icon to zoom in on the camera and the façade.

14. Select the Camera, right-click, and select **Move** from the quad menu.

15. Hold down the SHIFT key and then drag the X-axis of the Move Gizmo to the right.

16. Release the mouse when it is positioned directly below the target point.

17. In the Clone Options dialog, be sure that Copy is selected in the Object area, change the Name to **Center - Camera Correction**, and then click OK (see Figure 5–14).

NOTE If we choose Instance here, any modifiers we apply to either camera will be applied to both. For more information on these options, refer to the "Clone" topic in Chapter 2 and search for "instance" and "reference" in the VIZr Glossary.

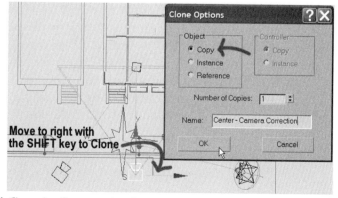

Figure 5–14 *Clone the Camera using the* SHIFT *key*

Make any adjustments that you wish to fine-tune the camera placement.

 NOTE Our camera is at X=13', Y=–30', Z=4', and its Target is at X=13', Y=4', Z=10'.

18. Minimize the Top viewport (using the Minimize Viewport icon or by pressing ALT + W again).

19. Right-click in the "Left – Looking Up" viewport to make it active.

20. Right-click on the viewport label, choose Views > Center - Camera Correction, and then Maximize the camera viewport.

Notice that it looks like our townhouse is "falling" out of the frame. We can fix this with a Camera Correction Modifier.

Make sure that the "Center – Camera Correction" Camera is still selected (look on the Modify Panel). Use the Selection Floater to reselect it if necessary.

21. Right-click in the viewport and choose **Apply Camera Correction Modifier** (see Figure 5–15).

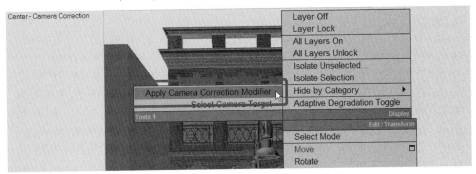

Figure 5–15 *Apply a Camera Correction Modifier to the selected Camera*

Camera Correction converts a 3-point perspective into a 2-point perspective. Two-point perspectives show vertical elements as vertical rather than converging towards a third point. For more information, search for "Camera correction modifier" under Help. When you apply the Camera correction modifier, it initially takes a "guess" at what values are necessary. If you modify the camera view later, you may need to adjust those settings in the Modify panel for the selected camera.

Direct your attention to the lamppost. It is a bit too far to the left in the composition. Combined with the front door, it puts too much emphasis on the left side. Let's adjust the camera angle slightly using the Viewport Navigation Tools.

22. Use the Truck Mode (pan hand) to pan the Camera to position the lamppost more to the right (see Figure 5–16).

 NOTE Our Camera and Target ended up at X=9′, Y= –30′, Z=4′, and its Target is at X=9′, Y=4′, Z=10′.

Figure 5–16 *Truck the Camera to fine-tune the Camera Viewport view*

23. Quick Render the scene again (see Render Stats 05.03).

24. Save the file.

 Render Stats 05.03

Render Time: 42 seconds
Radiosity Processing Time: NA
Music Selection: "Young Americans [Single Version]" David Bowie – Best of Bowie [Virgin US/Canada] (2002)
Color Plate: NA
JPG File: Rendering 05.03.jpg

Adding Entourage

Let's make a couple more changes to dress up the scene a bit. We can add a tree to the planting area to throw some shadow play onto the sidewalk and create a more interesting scene.

25. On the keyboard, press the **T** key to set the Top viewport current (and maximized).

26. From the EASYnat menu choose **Create**.

NOTE You must have EASYnat installed to have this menu. You can follow the link in the Content Browser to download and install the free plugin. Please refer to the 3rd Party Plugins section in Chapter 2 for more information.

If you do not have EASYnat, and do not wish to install it, please skip down to the "Apply Smoothing" topic below to continue without the tree.

27. Select **Horsechestnut-a** from the scrolling icon list.

28. In the Plant Morphology area, set the Age to **20**.

    Notice that the Diameter and Height adjust automatically based on the age chosen for the tree. Conversely, if you adjust these two settings, it will affect the Age of the tree.

29. Drag the Season slider to choose **Summer**.

30. Set the Plant Tuning to **2D**.

We will use the 2D version to optimize for speed. This will help us as we fine-tune settings and make adjustments. Once we get the placement and sizing right we will change it to Hybrid for rendering. The Hybrid option is a combination of 3D geometry and 2D geometry called Billboards. For more information, view the Help file accessed via the EASYnat pulldown after the plugin has been installed

31. Click the Generate tree button to create the tree, and then click the Close button to exit the dialog (see Figure 5–17).

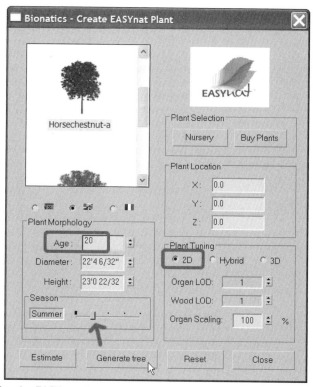

Figure 5–17 *Use the EASYnat plugin to create a tree for our scene*

The tree appears in Top view as a simple cross. It will be placed at the origin (the lower-left corner of the townhouse façade) and we will need to move it into place. The tree will remain selected after closing the dialog.

32. Right-click and choose **Move** from the Edit/Transform quad menu.

33. Move the tree so that it is between the two lampposts on the right (see Figure 5–18).

 NOTE We placed our tree at: X=25′, Y=−18′, Z=0′. We arrived at this position through trial and error with the goal of having a bit of the tree "peak" into the right side of the rendering.

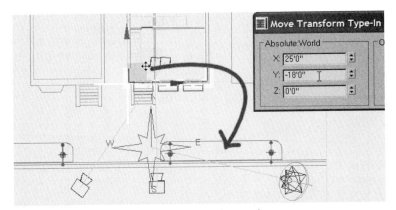

Figure 5–18 *Move the tree into position between the two lampposts*

A 2D tree is made up of two faces set perpendicular to each other. In order to get an idea about the shadow the final tree will cast, it is a good idea to rotate the tree so that one of its faces is perpendicular to the sun Daylight System.

34. Right-click and choose **Rotate** from the quad menu.

35. Use the Z constraint to Rotate the tree **–15°** (see Figure 5–19).

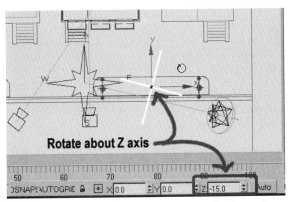

Figure 5–19 *Rotate the tree about –15° to make it perpendicular to the sun*

Now that we have our tree positioned, let's see how it looks in the rendered scene.

36. Type **C** to set the camera view active, and then click the Quick Render icon for another test rendering (see Render Stats 05.04).

Render Stats 05.04

Render Time: 49 seconds
Radiosity Processing Time: NA
Music Selection: "In the End" Linkin Park – Hybrid Theory (2000)
Color Plate: NA
JPG File: Rendering 05.04.jpg

You should have some nice shadow play from the balconies streaming across the façade while the tree's leaves cast a shadow on the sidewalk in the foreground. The tree should be peeking into the scene from the right. The rendering is shaping up nicely; however, from this camera angle the lamppost in the foreground appears much too segmented. We will adjust this with smoothing in the next topic.

Apply Smoothing

We can easily smooth out the segmented look on the lamppost. We use a tool on the modifier panel to apply smoothing to the lamppost. Based on the parameters set for smoothing, VIZr decides which surface intersections should render smooth (curved) and which ought to be corners. When you turn smoothing on, the angle designates the threshold of smoothing. When the angle between the face normals of two surfaces is smaller than the smoothing angle, the surfaces will smooth (see Figure 5–20). A face normal is an imaginary line (vector) pointing away from a surface at a 90° angle. The positive direction of this vector is considered the "front" of the surface, while the negative is the back.

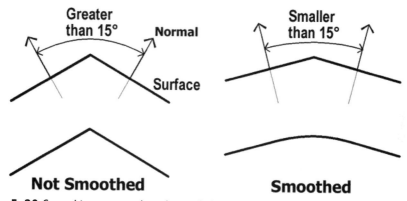

Figure 5–20 *Smoothing occurs when the angle between face normals is smaller than the Smooth-angle setting*

37. Click on the lamp post base closest to us in the foreground.

Notice that each separate piece of the lamppost is selectable. This is useful because we can Smooth only those pieces that need it and which are closest to the Camera.

38. In the Modify panel, click the Smooth button.

39. Adjust the Threshold setting to **50**.

40. Click the Quick Render icon again for another test rendering (see Render Stats 05.05).

With these modifications, we are starting to get a good sense of what the final rendering will be like.

Render Stats 05.05

Render Time: 53 seconds
Radiosity Processing Time: NA
Music Selection: "Beth" Kiss – The Very Best of Kiss (2002)
Color Plate: NA
JPG File: Rendering 05.05.jpg

41. Save the file.

Create Another Camera

Let's explore one more camera angle. We will try to make a more dramatic and closer shot of the façade. For this we will frame the whole façade and then use the Crop option to set the section for the output.

42. On the keyboard, press the **T** key to set the Top viewport current again.

43. Select the "Center – Camera Correction" Camera, right-click, and then choose **Move** from the quad menu.

44. As before, hold down the SHIFT key and move the camera with the XY-axis handle on the Move Gizmo up and to the right, just above the tree that we inserted earlier.

45. In the Clone Options dialog, set the Object to Copy.

46. Change the name to **Right - Cropped Render**, and then click OK to exit the dialog (see Figure 5–21).

 NOTE Our camera ended up at X=25', Y=–14', Z=4'.

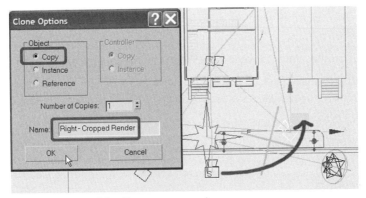

Figure 5–21 Clone a copy of the Camera next to the tree

The new Camera will remain selected.

47. Right-click and choose **Select Camera Target**.

    Be sure that Move mode is active. If it isn't, right-click and choose Move from the Edit/Transform quad menu.

48. Using the XY-axis handle on the Move Gizmo, Move the Camera Target between the center balcony railings of the façade (see Figure 5–22).

 NOTE Our Target ended up at X=12′, Y=−1′, Z=10′.

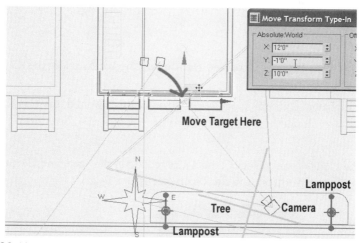

Figure 5–22 *Move the Camera Target between the Balcony Railings*

49. On the keyboard, press the **C** key to switch to the camera viewport.

50. Right-click on the viewport label, and choose **Views > Right – Cropped Render** to view through the new Camera.

Making Camera Adjustments

We will still need to make some adjustments to this view. For instance, we may need to adjust the Camera position so it's not so low. This camera angle may "feel" uncomfortable. We will also need to narrow the view of this shot to make the effect more dramatic. We will do this by adjusting the Height output size.

51. On the Rendering toolbar, click the Render Scene icon (or choose **Render** from the Rendering menu).

 NOTE Don't click Quick Render this time. We need to make adjustments in the Render dialog, we are not actually ready to create another rendering yet.

52. In the Output Size area, set the Width to **300**, and then close the Render Scene dialog without rendering yet.

This makes the view in our viewport *very* dramatic. However, this is not quite the look that we desired. Let's apply some final tweaks to the Camera to address this.

53. Open the Selection Floater (see the "Selection Floater" topic in Chapter 2). In the List Types area, click the None button to deselect everything.

54. Place a checkmark in only the Cameras checkbox.

 TIP This approach is useful in finding an object of a particular type and selecting quickly rather than scrolling through the entire list.

55. Double-click **Right – Cropped Render** on the left to select this Camera and exit the dialog.

You will notice that a Camera Correction Modifier is applied on this Camera. This is because we cloned it from the "Center – Camera Correction" camera, above.

56. In the Modify panel, select Camera Correction in the stack and click the Remove modifier from the stack icon (see Figure 5–23).

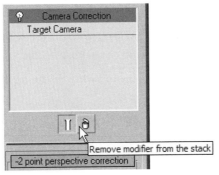

Figure 5–23 *Remove the Camera Correction Modifier from the Right – CroppedRender Camera*

57. In the Parameters rollout of the Modify panel, click the 35mm button.

This will adjust the field of view of the Camera to zoom-in closer on the façade of the townhouse only.

58. Click the Orbit Camera icon and orbit the camera so that the sides of the townhouse appear straight up and down.

59. Click the Truck Mode icon and truck the camera up (move the mouse in a downward motion) to be more at the level of the second floor windows.

Make additional adjustments to make the viewport match Figure 5–24.

 NOTE Our Camera and Target ended up at X=25′, Y=–16′, Z=22′, and X=14′, Y=1′, Z=22′ respectively.

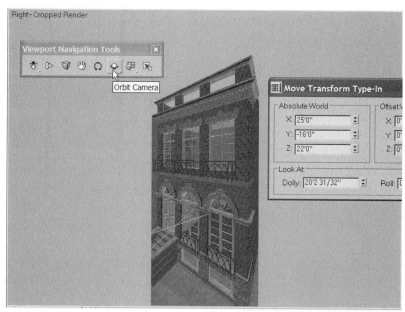

Figure 5–24 *Use the Viewport Navigation controls to adjust the Camera view*

We're ready for another test render.

60. Click the Quick Render icon (see Render Stats 05.06).

 Render Stats 05.06

Render Time: 27 seconds
Radiosity Processing Time: NA
Music Selection: "Fotografía" Juanes; Nelly Furtado – Un Dia Normal (2002)
Color Plate: NA
JPG File: Rendering 05.06.jpg

61. Save the file.

This rendering is a bit too vertical. We will address this below before the next test render. However, now that we are up close to the façade, perhaps we might benefit from making some material adjustments.

MATERIAL MODIFICATIONS

Let's dress up the model a bit with a couple of material changes. While you are working in VIZr, you may wish to experiment with a different material than the one assigned. You can easily find the material in question on the palette and then swap it for another.

New Front Door Material

Let's liven up the front door a bit with some new paint.

1. Click on the Scene – Unused palette tab.

2. Right-click on the Create New material icon and choose **New > Paint Semi–gloss** (see Figure 5–25).

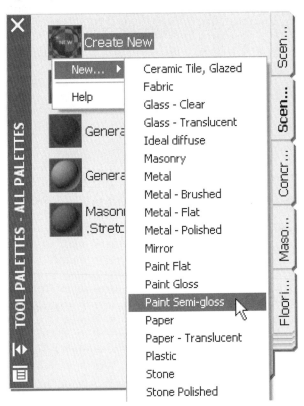

Figure 5–25 *Create a new Material from the Scene – Unused palette*

3. Scroll to the bottom of the list, right-click on Paint – Semi Gloss, and choose **Properties**.

4. A the top of the Material editor, beneath the preview images, change the name (in the dropdown list) to **MVIZr - Front Door**.

5. Click on the Diffuse Color swatch and set Red: **120**, Green: **20**, and Blue: **20**.

6. Drag the sample sphere to the scene and drop it on top of the front door (see Figure 5–26).

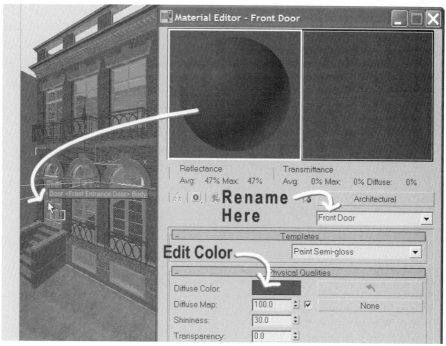

Figure 5–26 *Create a Red Door Paint Material*

Before you drop it, wait for the Door *<Front Entrance Door>* Body tool tip to appear. This indicates that the material will be applied to that object. However, notice that this applies the material only to the raised panels (you may recall that that the door stiles and panels were separate objects back in Chapter3).

7. Repeat this step to drag the material to the Door Panel as well: *Door <Front Entrance Door> Door Panel*.

8. Close the Material Editor and then Save the file.

CREATE BRICK ARCHES

It might make a more interesting rendering if we change the material applied to the arches on the first floor. Let's see how they look in brick rather than stone.

Create a New Material

To start, we need to create a new material for bricks.

1. On the Scene – Unused palette, right-click the New icon and choose **New > Stone**.

    This will create a new material based on the Stone template.

2. Scroll all the way to the bottom of the Scene – Unused palette, right-click the Stone material, and choose **Properties**.

3. Rename the material **MVIZr – Brick Arch**.

4. Next to the Diffuse Map slot, click the map button (currently labeled "None"—see Figure 5–27).

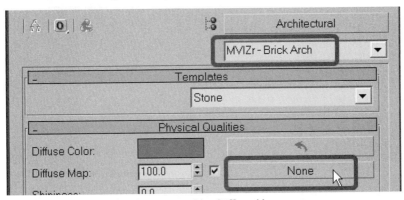

Figure 5–27 *Click the Map Slot button to add a Diffuse Map*

5. In the Material/Map Browser, double-click Bitmap item at the top of the list.

6. Browse to the *C:\MasterVIZr\Maps* folder, choose *MVIZr_BrickArch.jpg,* and then click Open.

7. Just beneath the preview sphere at the top of the Material Editor, click the Go to parent icon.

    This will return you to the top of the material tree.

8. Click the curved arrow button above the diffuse map slot.

Notice that the color swatch for Diffuse color changes to a shade of brownish-red. This curved arrow button sets the Diffuse color to match the average color of the bitmap image that we just loaded.

Apply Brick Arch Material

9. Open the Selection Floater.

If List Types are still filtered to Cameras only, click the All button to show all object types in the list.

10. In the text field at the top, type "**window <first floor transom> b**."

 NOTE Watch the change in the selection in the list as you type. By typing in part of the name, you can find objects from a very long list.

11. Click the Select button to select the three highlighted arches (see Figure 5–28).

Figure 5–28 *Type in a partial name in the Selection Floater to quickly select objects with similar names*

12. Right-click on the MVIZr – Brick Arch material (on the Scene – Unused palette) and choose **Apply to Selected**.

Adjust Map Position

Notice that the map is not positioned quite right. We need to make some modifications to the way that this map is applied to the surface of the arch.

If you closed the Material Editor, right-click the MVIZr – Brick Arch material (which is now on the Scene – In Use palette) and choose **Properties**.

13. On the Material Editor, expand the Texture Scaling rollout.

14. Change the Width to **8'-8"** and the Height to **4'-4"**.

TIP While you work, it might be easier to work in Front view. Press the F key on the keyboard, and then right-click the Front viewport label and choose Smooth + Highlight.

These sizes tell VIZr how big the map is that we loaded. If you were to view the image (which you could do by clicking on the map slot button, and then clicking the View Image button—be sure to click Go to Parent to return to the top of the Material Editor if you do this), you would see that this image represents 8′-8″ x 4′-4″ of brick.

15. Change the U Offset to **4′-4″** and the V Offset to **2′-9″**.

To understand these settings, you can imagine that the bitmap image has an insertion point (like an ADT block). The insertion point for the bitmap is its lower-left corner. The insertion point of the brick arch object, however, is at the center of the arch. Therefore, our map is off by half each direction. When you position a map, the letters U V W are used in place of X Y Z. The map is 2D, and it is applied to the vertical surface of the arch. So in this case, the X position is controlled by the U Offset, and the V Offset shifts the Z position (the vertical face of the arch). We need to move the map to the right by half the width (which is the difference between the insertion point of the map and the arch). If we move the V Offset 2′-9″, it aligns the bricks up with the arch nicely in the vertical position. Again, it will be easiest to see this in a front view (see Figure 5–29).

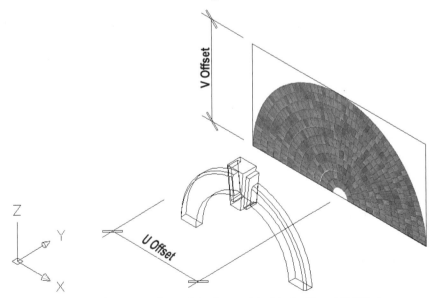

Figure 5–29 *Align the bitmap to the arch in the model with the U and V Offsets*

The only problem that we still have is that the entire arch, keystone and all, has this new material applied. This is because the mass elements that make up the arch and the keystone are all the same style. To correct this we need to go back to ADT.

    16. Save the scene.

Drag the New Material into ADT

We can use this new material that we made in ADT as well as in VIZr. This will also help us get it properly assigned to the correct part of the arch.

    For this sequence, you will need both ADT and VIZr running. If you closed ADT, please reopen it now.

    17. From Project Navigator, on the Constructs tab, double-click to open the *First Floor Construct*.

    Switch to VIZr and position its window so that you can see both VIZr and ADT onscreen at the same time.

    18. From the Scene – In Use palette, drag the MVIZr – Brick Arch material from VIZr and drop it into the ADT drawing window (see Figure 5–30).

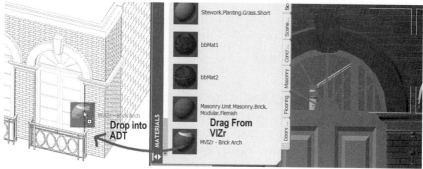

Figure 5–30 *Drag a Material from VIZr to ADT*

This will drop the material we created in VIZr into ADT as a Render Material that can be associated with an ADT Material Definition. We will create this Material Definition next.

    Leave VIZr running, but switch to ADT.

    19. From the Format menu, choose **Material Definitions**.

    The Style Manager will open filtered to the Material Definitions node.

    20. Right-click on Masonry.Unit Masonry.Brick.Modular.Running and choose **Copy**. Right-click again and choose **Paste**.

    21. Rename the new material **MVIZr – Brick Arch**.

22. Double-click MVIZr – Brick Arch to edit it and then click the Display
    Properties tab.

23. Double-click General beneath the Display Representations column to edit it.

24. Click the Other tab, and then from the Render Material list, choose
    MVIZr – Brick Arch, and then click OK (see Figure 5–31).

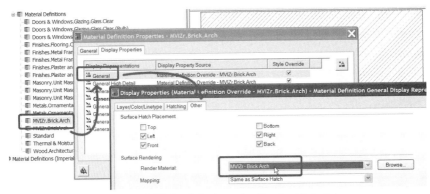

Figure 5–31 *Choose the new MVIZr.Brick.Arch as the Render Material*

25. Repeat this process for each of the Display Representations on the list
    except General Low Detail.

26. Click OK twice when finished to return to the drawing window, and then
    save the file.

Since we have linked to the High Detail Display Configuration at the start of this
chapter, you really only need to be concerned with the General High Detail Display
Representation. However, it is good practice to do them all since you may
want to reuse this material definition again in the future on other ADT models.

Edit the Half Round Arch

You will not see the effect of these edits yet, since we have not yet applied the material
to the arch in the model. The Arch as been applied to the Door/Window
Assembly as a Display Block (see Chapter 4), so we will need to insert it as a
standalone block and then edit it in place. Each display block will receive a single
material in VIZr. A simple remedy to this situation is to create a second display
block for the Window—one will contain just the arch, while the other contains
just the keystone.

27. From the Insert menu, choose **Block**.

28. From the Name list, choose *Half Round Arch*, be sure that Specify On-screen
    is chosen for Insertion Point, but not for Scale and Rotation, and then
    click OK.

29. Click anywhere off to the side of the model to place the block.

30. Double-click on the block to edit it in place. Click OK to confirm in the Reference Edit dialog.

31. Select the Keystone Mass Element, right-click, and choose **Copy Mass Element Style and Assign**.

32. On the General tab, change the name to **Keystone** and then click OK.

33. Select the arch Mass Element, right-click, and choose **Edit Mass Element Style**.

34. Click on the Materials tab.

35. Choose **MVIZr – Brick Arch** from the Material Definition list next to the Body Component (see Figure 5–32).

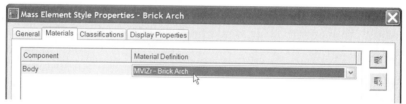

Figure 5–32 *Assign the new MVIZr – Brick Arch Material to the Arch Mass Element Style*

36. Click OK to return to the drawing, and then, on the Refedit toolbar, click the Save icon.

Notice that the brick pattern that appears looks like normal non-arched bricks. This is because in wireframe and hidden line we see a hatch pattern, not the render material. If you wish to see arched bricks in these display modes, you will need to create a custom hatch pattern. Creating custom hatch patterns is beyond the scope of this book.

37. Erase the *Half Round Arch* block.

It is no longer needed, since the change applies to the arches on the Windows as well.

Display the Texture in ADT

To see the texture in ADT, we must add a path to the *Maps* folder.

38. From the Format menu, choose **Options**.

39. On the Files tab, highlight the Textures item and then click the Add button.

40. Click the Browse button and then browse to the *C:\MasterVIZr\Maps* folder and add the path. Click OK to dismiss the Options dialog (see Figure 5–33).

Figure 5–33 *Add a Path to the MasterVIZr\Maps folder*

41. Save and Close the *First Floor Construct* file and then reopen it from the Project Navigator.

42. From the View menu, choose **Shade > Gouraud Shaded**.

Closing and reopening the file will force it to re-read the texture paths. If this does not work, quit ADT and then re-launch it. The texture may still require some adjustment to show in the correct location in ADT as well. Remember the edits that we made to the offsets above. The same would be required in ADT. For now we will remain focused on the VIZr side, and later, if you wish, you can return to ADT and adjust the mapping. This would be done in the Display Properties of the Material Definition, on the Hatch tab using the X and Y Offsets.

43. Save and Close the *First Floor Construct*.

Reload the ADT Model in VIZr

Switch to VIZ Render. Do not start a new session of VIZr. Do not use the Link to VIZ Render option in ADT.

44. On the File Link toolbar, click the **Use scene material definitions** toggle icon.

 NOTE See the "Reload From VIZr" topic in Chapter 2 for more information on reload options.

45. On the File Link toolbar, click the Reload geometry and materials icon.

Notice that we have lost the material assignment on the front door. This is because we did not toggle the "Use scene material assignments (on Reload)" setting on. Since we have made material modifications in the ADT file that we wanted to use here, this was necessary. However, if you recall from Chapter 2, our MVIZr - Front Door material will still be present on the Scene – Unused palette in this scene. All we need to do is re-apply it to the door from there.

46. From the Scene – Unused palette, assign the MVIZr - Front Door to the front door again.

HOW THE BRICK BITMAP WAS CREATED

We have provided the image file for your brick arch here. It is a JPG file that represents bricks in an arch pattern. You can create your own brick arch textures to use with this technique. To do so, photograph an arch that matches the one that you like, or use your favorite image editing program like Adobe Photoshop or Paint Shop Pro to paint your own brick arch JPG file. If you have access to Autodesk VIZ or 3ds max, these programs provide advanced tools that make it easier to model an arch such as this that are not available in VIZ Render or ADT.

47. Save the scene.

PREPARE THE FINAL RENDERING

Let's see how all of these material modifications are affecting our rendered scene. Before we configure our final render settings, let's do one more test render.

Apply a Crop to the Rendering

The last rendering was a bit too tall. We can frame the rendering using the Crop option.

1. Minimize the Front viewport and set the Right – Cropped Render Camera viewport active.

2. On the Rendering toolbar, select **Crop** from the Render Type list at the right (see Figure 5–34).

Figure 5–34 *Choose the Crop option before clicking Quick Render*

3. Click the Quick Render icon.

This time you will notice that the rendering does not start automatically. This is because you are being prompted to select the area to be cropped. (Look in the Status bar at the bottom left – see Figure 2-1)

4. Resize the crop boundary as shown in Figure 5–35 and then click the OK button at the bottom-right corner of the viewport (see Render Stats 05.07).

Figure 5–35 *Quick Render this time makes a Crop Boundary box appear*

 Render Stats 05.07

Render Time: 42 seconds
Radiosity Processing Time: NA
Music Selection: "MacArthur Park" Donna Summer – On the Radio - Greatest Hits
Vols. I & II (1987)
Color Plate: NA
JPG File: Rendering 05.07.jpg

We are ready to render the final shot. We will be using this last camera. However, feel free to make final renderings from any of the other Cameras that we created in this chapter. Set your favorite camera to active after adjusting the following settings.

Adjust Render Parameters

Since we will be using Radiosity in the final rendering, we need to make some adjustments to some of the objects in our scene that do not need to have the same meshing properties as the rest of the scene. This will help speed the processing time.

For instance, when we render the "Right – Cropped Render" Camera view, the sidewalk, street, lamppost, grass area, and the building adjacent to the townhouse on the right are not visible. Therefore, in order to save unnecessary Radiosity processing time, we will modify the Radiosity parameters for these items.

5. Right-click in the Top viewport to make it current.

6. Click in the lower-right corner and drag up and to the left to create a crossing selection surrounding the lamppost and grass area.

   This will select the street, sidewalk, lampposts, and the grass areas. The Daylight System and Cameras may also be selected. We can remove them from the selection set via the Selection Floater.

7. Open the Selection Floater. In the List Types area, click the All button, and then deselect the Geometry checkbox.

8. Beneath the list of objects, click the None button to deselect all items.

9. On the right, in the List Types area, click the Invert button.

   All of the geometry selected onscreen should still be highlighted here.

10. Click the Select button and then the Close button to dismiss the Selection Floater.

    Let's add the neighboring buildings to the selection set as well.

11. Hold down the CTRL key and select each of the neighboring buildings.

   TIP Try clicking them at the entrance stairs for easier selection.

12. On the Selection toolbar, click in the Named Selection Sets dropdown list, type **Radiosity Exclusions** for a name, and then press ENTER.

Remembering to save your complex selections will save you time later should you need to restore them.

NOTE For more information on Saved Selections, see the "Selection Floater" heading in Chapter 2.

13. Right-click and choose **Rendering Properties**.

14. In the Geometric Object Radiosity Properties area, remove the checkmark from the Use Global Subdivision Settings checkbox.

    You may need to click twice depending on whether the settings vary.

15. Also uncheck the Subdivide checkbox and place a checkmark in the Exclude from Regathering checkbox (see Figure 5–36).

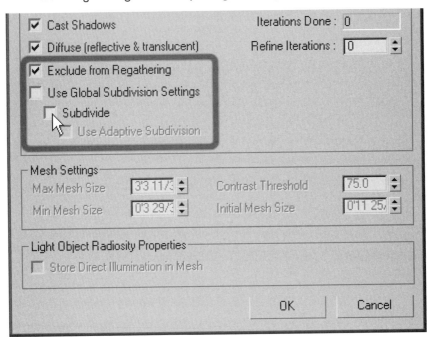

Figure 5–36 *Configure settings in the Render Properties dialog*

16. Click **OK** to accept the changes and to close the dialog.

17. Click to the side of the scene (not on an object) to clear the current selection set (or use the Selection Floater to select "None") and then Save the scene.

Produce the Final Rendering

We are finally ready to produce our final rendering. We just need to open the render dialog and adjust some parameters, and we are ready to go.

18. Make the **Right – Cropped Render** camera viewport active.

19. On the Rendering toolbar, click the Render Scene icon.

20. In the Render Options rollout, place a checkmark in the Compute Radiosity when Required checkbox.

This setting, if checked, will automatically re-calculate Radiosity if some change to the scene invalidates the existing solution. For example, adding a Camera would not affect Radiosity where adding a Light would. For more information, refer to the following section: Help > Rendering > Rendering Rollouts > Render Options Rollout.

21. Click on the Radiosity tab, and in the Radiosity Meshing Parameters rollout, place a checkmark in the Enabled checkbox and then set the Maximum Mesh Size to: **2′0″** (see Figure 5–37).

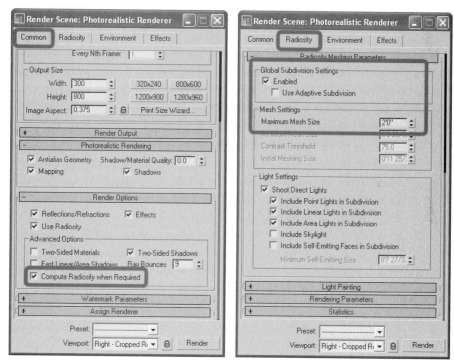

Figure 5–37 *Turn on Radiosity and configure the Meshing size for the final rendering*

22. Without closing the Render Scene dialog, choose **Save** from the File menu.

23. Verify that the "Right – Cropped Render" Camera is showing in the viewport list at the bottom of the dialog, and then click the Render button.

24. Adjust the cropping area if necessary and then click the OK button at the bottom-right of the camera viewport (see Render Stats 05.08).

Render Stats 05.08

Render Time: 12 minutes 58 seconds
Radiosity Processing Time: 7 minutes 18 seconds
Music Selection: "The Dark Knight" Branford Marsalis Quartet – Crazy People Music
Color Plate: C–5
JPG File: Rendering 05.08.jpg

ADDITIONAL EXERCISES

If you wish to go further, try rendering another camera angle; you may want to adjust the meshing size parameters for the objects in the Radiosity Exclusion selection set saved above. Even though they are visible in those Camera views, they are not the focal objects in the scene. Therefore, try a setting 4 to 5 times what we chose above. (In other words, try a setting of 8′ to 10′ since we set the meshing for all the included objects to 2′ above.) Also, we may want to change our Bionatics tree to Hybrid or 3D to give the best results in the final rendering. To do this, simply select the tree object and then change its parameters on the Modify panel. A word of caution: the combination of older aged plants and 3D Plant Tuning can increase the amount of faces in your scene. Increased face count is to be avoided whenever possible because it will slow your system, as well as increase your rendering times. So, go with 2D or Hybrid unless it is imperative to render near photo-quality entourage.

To complete the high resolution rendering you see in the color plates, load the *Ch05_Final Render_High Res.rps*.

CONCLUSION AND SUMMARY

In this chapter we combined the use of several different Camera angles with judicious configuration of rendering and Radiosity settings to arrive at a high quality final result. In this chapter we learned the following:

How to set up a View drawing in Project Navigator within ADT for the specific purpose of visualizing the design, and how to Link this file to VIZr.

How to Create and Manipulate Cameras to set up the shot that you wish to render.

Copy an existing Camera and modify it to easily create a new Camera view.

Add accoutrement to generate believability and to frame your renderings.

Brick Arches can be achieved by first painting or photographing a bitmap image of the arch, and then applying this to a material.

Careful configuration of settings can help to generate a Radiosity solution that adds more believability with little extra rendering time.

CHAPTER 6

Interior Rendering with Daylight

INTRODUCTION

In this chapter, we will explore the creation of an interior daylight rendering. We will be working in a small dataset comprised of a living room with a view to the outside. Once again we will light our scene using the Daylight System. However, since we are creating an interior this time, we will also supplement the daylight with some artificial lighting. A large focus of this chapter will be on materials and their impact on a scene. We will use out-of-the-box materials as well as custom-created ones, and those that populate our scene as a result of using manufacturer's i-drop content.

OBJECTIVES

The primary objective of this chapter is to achieve a high quality interior rendering lit with daylight. Many other goals and objectives will be the subject of this chapter as well. These include the following:

- Understanding and using Substitution
- Working with Interior Daylight
- Creating and manipulating Materials
- Accessing items from the Content Browser

THE LIVING ROOM DATASET

The model for this chapter's project is a small living room scene. Basic ADT objects like Walls, Door, and Windows were used to construct the scene. A Space object makes up the floor and ceiling. The Window blinds were modeled from a Curtain Wall object where mullions were used for the blades of the blinds. The main light in the space will be generated by daylight coming in from the windows. We will also have some supplemental lighting in the interior in the form of recessed can lights and candles. The model contains some Architectural Desktop Multi-View Block objects for the furniture and other fixtures in the space. We will use the Substitution modifier to replace these with more detailed content generated outside ADT and acquired from various manufacturers' web sites. All work in this chapter will take place in VIZr.

Start VIZ Render and Load the Living Room Dataset

1. If you have not already done so, install the dataset files located on the Mastering VIZ Render CD ROM.

   Refer to "Files Included on the CD ROM" in the Preface for information on installing the sample files included on the CD.

2. Launch VIZ Render from the icon on your desktop.

   You can also launch VIZr Render by choosing the appropriate item from your Windows Start menu in the Autodesk group. In Windows XP, look under "All Programs"; in Windows 2000, look under "Programs." Be sure to load VIZ Render, and not Architectural Desktop, for this exercise.

3. From the File menu, choose **Open**.

4. Click to open the folder list, and choose *My Computer* and then your C: drive.

5. Double-click on the *MasterVIZr* folder, and then the *Chapter06* folder.

6. Double-click *Living Room.drf* to open the file.

SUBSTITUTION

When you build your models in ADT, it is not always practical to build all items to the level of detail required for high-quality rendering. The needs of the ADT model often do not require the use of highly detailed or specific 3D artifacts. For instance, if you place furniture in an ADT model that is being used primarily for construction document production, it is often sufficient to place very simple abstract furniture symbols. However, when you link this model to VIZr, these abstracted items (often ADT Multi-View Blocks or AutoCAD Blocks) are usually too simple and generic to produce high-quality renderings or convey specific design intent. In fact, these Content items may actually only be 2D symbols that have no 3D characteristics at all! This would make a very boring rendering indeed. To remedy this situation, we can use the Substitution tool in VIZ Render. With this utility, we can swap a more detailed and appropriate 3D artifact for the simplified "place-holder" object contained in the ADT model.

Replacing ADT MVBlocks via Substitution

1. In the Top viewport, select the longer couch (running horizontally—see item 1 in Figure 6–1).

2. From the Utilities panel, choose Substitute Manager (see item 3 in Figure 6–1).

 NOTE The Utilities Panel is a tab (right next to the Modify Panel tab—see item 2 in Figure 6–2) on the Command Panel which is usually docked vertically on the right side of the screen.

3. In the Substitution rollout, click the Create Substitute button (see item 4 in Figure 6–1).

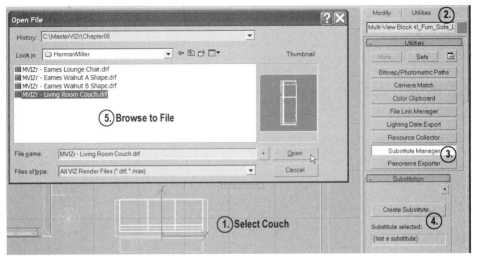

Figure 6–1 *Open the Substitute Manager to swap in a more detailed couch than the stand-in one from ADT*

4. In the Open File dialog, browse to the *C:/MasterVIZr/Chapter06* folder and double-click on *MVIZr - Living Room Couch.drf* to open it (see item 5 in Figure 6–1).

 NOTE If you completed Chapter 2, this file is exactly like the one that you created there. It was created from the i-drop content on Herman Miller's web site. Please refer to the "i-drop" topic in Chapter 2 for more information.

Repeat the process for the other pieces of furniture.

5. Select the round side table next, and then substitute it with the *MVIZr - Eames Walnut A Shape.drf* in the same folder.

6. Substitute the file named *MVIZr - Eames Walnut B Shape.drf* for the square side table.

NOTE The MVIZr - Eames Walnut B Shape.drf table substitution is actually round, not square.

7. Using the CTRL key, select both of the side chairs near the window.

 NOTE Notice that these chairs do not show up in the other views. This is because in ADT, these are simple 2D blocks, as we noted at the beginning of this chapter.

8. In the Substitution rollout, click the Create Substitute button again and browse to the *C:/MasterVIZr/Chapter06* folder again.

9. Choose *MVIZr - AaltoLounge.drf* and then click Open to substitute both chairs with this item at the same time (see Figure 6–2).

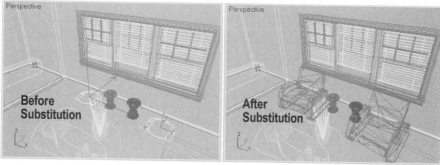

Figure 6–2 *Replace simple 2D place-holder blocks with a detailed 3D Chair*

We have just one more item to substitute—the loveseat. We will have to work just a bit harder to get the substituted object oriented properly however.

10. Select the loveseat, and then substitute it with the *MVIZr - Eames Lounge Chair.drf* (see Figure 6–3).

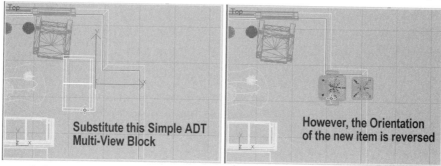

Figure 6–3 *The substitute for the love seat is oriented incorrectly*

Reorient the Substituted Item

As you can see, the substituted item comes in oriented the opposite way. However, since the MVIZr - Eames Lounge Chair is actually two separate objects, if you try to rotate it, each object will rotate separately around its own pivot point. What we need to do is instruct the Rotate transform to operate around a common pivot point instead.

11. On the toolbars, right-click on an empty gray space next to any toolbar. From the menu that appears, choose Transforms (see Figure 2–2 in Chapter 2 for a similar example).

    Be sure that the Select and Rotate icon is depressed on the Transforms toolbar. If it is not, click it now.

12. Hold down the CTRL key, and select both the chair and the ottoman.

13. On the Transforms toolbar, click and hold down the icon directly to the right of the Reference Coordinate System dropdown list to make the flyout appear.

14. Choose Use Selection Center if it is not already selected (see Figure 6–4).

    The transform axis should appear in the center, between the chair and ottoman.

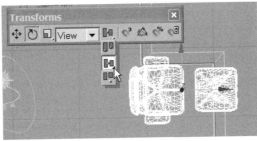

Figure 6–4 *Select both the chair and the ottoman and choose Reference Coordinate Center as the transform pivot*

15. Rotate the selection **180°**.

TIP Use the Angle Snap Toggle icon on the Transforms toolbar to rotate onscreen with precision. Right-click this icon to set the Snap Angle Increment.

16. On the Transforms toolbar, click the Select and Move icon.

17. Move the selection down slightly and to the left, as shown in Figure 6–5.

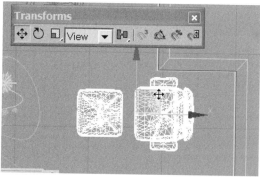

Figure 6–5 *Move the chair and ottoman selection to a more favorable position*

If you wish, dock the Transforms toolbar to keep it onscreen, or simply close it for now. You can always reload it later if necessary.

18. Save the scene.

INTERIOR DAYLIGHT

To create a dramatic scene, we want to position the daylight system so that the sun shines through the window and shows the shadows from the window blinds on the wall and the floor. To accomplish this, we will need to adjust a few settings for the daylight system as well as use the compass to help us position it properly.

Add a Daylight System

1. From the Create menu, choose **Daylight System**.

2. In the Daylight Object Creation dialog, choose **No**.

As we have seen before, the dialog that appears is essentially asking us if we are lighting an exterior scene. Since this is an interior scene, we want to answer "No" here.

3. For the center of the Compass, pick somewhere between the TV and the coffee table.

The exact location is not critical, this is just a convenient location.

4. Drag to size the compass large enough to be legible and complete the placement of the Daylight System's Orbital scale.

Recall the **Click – Drag – Release – Drag – Click** method covered in Chapter 1 if you need a refresher in creating a Daylight System.

5. On the Modify panel, rename the Daylight01 object to **Sun**.

Fine-Tune Daylight Position

The way that the model is constructed, the windows appear to be on the north side of the building (assuming of course that north is pointing in its traditional "up" position). It just so happens that these windows are actually facing to the south in the design and the model was simply built at an alternative orientation. Therefore, we need to adjust the Daylight System's orientation to reflect this.

6. On the Modify Panel, with the Daylight System still selected, expand the Control Parameters rollout (if it is not already) and then scroll all the way to the bottom of the panel.

7. In the Site area, set the Orbital Scale to **20'-0"**, and then change the North Direction parameter to **180**.

For this scene we want the light to stream in through the window and shed some light on the wall where the TV is. We will use the Front and Left viewports to help us position the sun just right to achieve this effect.

8. Be sure that the **Sun** object is still selected, and then adjust the Time to **14** Hours, **0** Mins. and **0** Secs.

9. Adjust the Month to **2**, leave the Day set to **21**, and the Year at **2005** (see Figure 6–6).

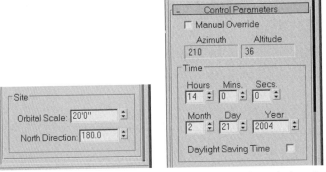

Figure 6–6 *Configure the settings for the Sun object to shine in through the windows*

Verify Sun Position

Now let's check to see how these changes match up to the effect we want in our scene.

10. Select the **Compass** object, right-click, and choose Move from the Edit\Transform Quad menu.

11. Slowly move the compass around in the scene and pay attention to the blue line connecting the Sun object to the Compass in each viewport (see Figure 6–7).

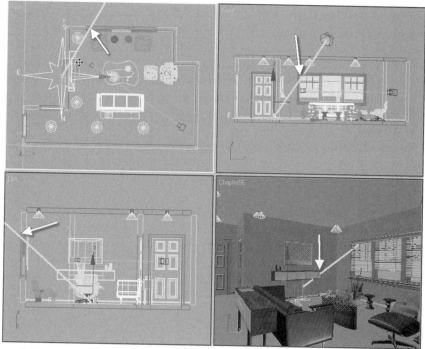

Figure 6–7 *Move the Target around the scene to verify that the Sun is at the desired angle*

As shown in Figure 6–7 above, what you are looking for is how the rays of sunlight will pass through the window. In the Top viewport, we have positioned the sun at the left side of the window. In the Left viewport we have positioned the sun at the top of the window. By doing this we can see in the Camera viewport where on the wall the sunlight will hit the wall from that position. If we continue to move the Daylight System around, we can get an idea as to what object the sun will be hitting at this time of day for anything in the scene. Feel free to continue to adjust the time and month settings to your liking.

12. Save the scene.

MATERIALS

In this topic we will concentrate on materials. Since we will be rendering quite a bit while testing materials we should adjust some settings to help speed up those test renderings before we begin.

Perform a Test Render

As we have discussed elsewhere, the Rays per Sample setting determines the number of rays of sunlight hitting any given point in your scene. A higher number will yield a better looking rendering but at the cost of longer rendering times. For our test renderings, let's reduce this number substantially from the default.

1. Select the Sun object.

2. On the Modify panel, under the IES Sky Parameters rollout, change the Rays per Sample to **5**.

3. On the Rendering toolbar, click the Render icon.

4. At the bottom of the dialog, choose **Load Preset** from the Preset list.

5. Browse to the *C:\MasterVIZr\Render Presets* folder, choose *CH06_Test Render*, and then click Open.

6. In the "Select Preset Categories" dialog, be sure that everything is selected and then click the Load button.

7. In the Render Scene dialog, click the Render button (see Render Stats 06.01).

 Render Stats 06.01

Render Time: 2 minutes 20 seconds
Radiosity Processing Time: 1 minute 30 seconds
Music Selection: "Panic" The Smiths – Singles
Color Plate: NA
JPG File: Rendering 06.01.jpg

If you study this rendering, one thing that is immediately apparent is the very "jaggy" edges around the TV and the TV console. Do not be concerned about this at this time. One of the settings that we have turned off by applying the Render Preset is anti-aliasing. Anti-aliasing is a kind of smoothing process that helps to diminish the "jaggies." Later, when we do a final rendering, we will turn on anti-aliasing.

THE WINDOW BLINDS

Let's start with the window blinds. They are constructed from a Curtain Wall object in ADT. This object has a simple out-of-the-box wood material applied to it. In this topic, we will make some modifications to this material.

Adjust Radiosity Settings

We will adjust the Radiosity settings for the material applied to the window blinds so that we can limit the amount of their tan color that is reflected back into the room.

1. In the Front viewport, select one of the window blinds, right-click, and choose **Query Material**.

   This will highlight the Doors & Windows.Wood Doors.Ash material on the Scene - In Use palette.

2. On the Scene - In Use palette, right-click on this material and choose **Properties**.

3. Change the name of the material to **MVIZr-Blinds**.

4. Expand the Adjust Radiosity rollout and change the Color Bleed Scale to **10**.

Reducing the value this way will reflect very little of the tan color of the blinds back into the scene. For detailed information on the settings that we are manipulating here, choose User Reference from the Help menu and click the Search tab. Search for "Adjust Radiosity Rollout."

5. Change the Indirect Bump Scale to **10** and the Reflectance Scale to **50**.

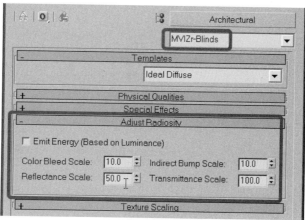

Figure 6–8 *Adjust the Radiosity parameters for the MVIZr-Blinds Material to reduce bleed and reflection*

The Indirect Bump Scale setting allows you to fine-tune a simulated bump map effect when the material is indirectly lit. Feel free to experiment with values here. Reducing the Reflectance scale limits the amount of light reflected back into the scene from a surface. Reducing this as we have will again help to limit the effect that the blinds have on the overall scene. Detailed information on both of these settings is also available in the same topic of the online help.

6. Save the scene.

7. On the Rendering toolbar, click the Quick Render icon (see Render Stats 06.02).

Render Stats 06.02

Render Time: 2 minutes 15 seconds
Radiosity Processing Time: 1 minute 30 seconds
Music Selection: "Tush" ZZ Top – Fandango
Color Plate: NA
JPG File: Rendering 06.02.jpg

WALLS

The Walls in this scene have a simple ADT drywall material applied to them. Let's make some adjustments to provide a bit more interest to the scene.

Create a Stucco Wall Material

The walls need a bit of color and texture to liven up the scene.

1. Select one of the walls, right-click, and choose **Query Material**.

   This will highlight the Finishes.Gypsum Board.Painted.White material on the Scene - In Use palette.

2. On the Scene - In Use palette, right-click on this material and choose **Properties**.

3. Change the name of the material to **MVIZr-Stucco Texture**.

4. Change the Diffuse Color to R: **154**, G: **144**, B: **116**.

5. Click the map button next to the Diffuse Map slot.

6. Click the Bitmap button, and in the Material/Map Browser, choose None and then click OK.

This will remove the bitmap assigned to the Diffuse channel and use the color values specified above instead. You will automatically be returned to the parent level of the Material Editor.

7. In the Adjust Radiosity rollout, change the Indirect Bump Scale to **20**.

8. In the Special Effects rollout, change the Bump setting to **40** (see Figure 6–9).

Figure 6–9 Increase the Bump effect for this material

Apply Bump Parameters

This will give the Walls more texture.

9. Click the Bump slot currently labeled Finishes.Gypsum Board.Painted.White.bump.

This will take us to the parameters of this map that is applied to the material to give the illusion of texture. Notice that this material currently uses a Bitmap (as indicated on the button next to the map name) to give this illusion. The name of this Bitmap is Finishes.Gypsum Board.Painted.White.bump.

10. Click the Bitmap button next to the name field.

11. In the Material/Map Browser dialog that appears, choose Noise, and then click OK (see Figure 6–10).

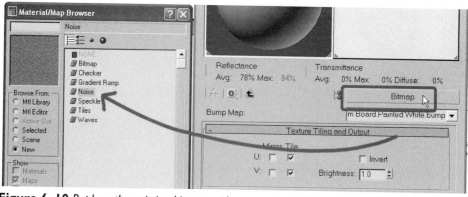

Figure 6–10 *Replace the existing bitmap with a procedural Noise map*

12. Change the name of the Noise map to **MVIZr-Bump-Noise-Stucco**.

13. Set the Noise Type to **Fractal**.

14. Set the Noise Threshold to High: **0.65** and Low: **0.45**.

The Noise types are really a matter of personal preference. Fractal tends to do a nice job of simulating believable randomness. A Noise map will include a range of gray shades in a random pattern from 0 (Black) to 255 (White). If you viewed the Noise map in section, it would appear as a series of hills and valleys. When used in the Bump channel it helps to simulate a roughness in the surface texture. The threshold settings used here are basically trimming off the tops and valleys of the noise map. This simulates a troweled stucco texture (see Figure 6–11).

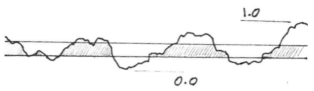

Figure 6–11 *A conceptual diagram of the Noise Bump Map effect*

In the sample sphere nothing seems to happen. Let's make a temporary change that will help us visualize the result of our changes without having to spend the time rendering.

15. Click the Go to Parent icon to return to the top of the material tree (see Figure 6–12).

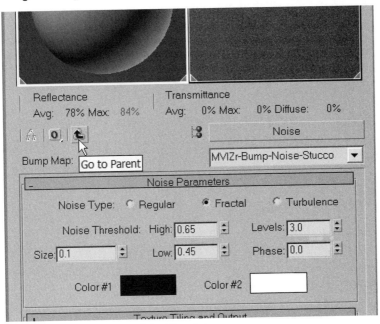

Figure 6–12 *Return to the Parent of the Material tree*

16. Expand the Texture Scaling rollout and then change the Width and the Height to **100′**.

17. Click on the Bump slot again to return to the Bump controls.

18. Change the Levels to **10** to add some more detail.

Now that you can see the effects of the texture, make any additional adjustments to the Threshold and Levels settings as desired.

19. Go to Parent, and in the Texture Scaling rollout, set the Width and Height to **1′-0″**.

If you want more information on these settings, search for Noise Map in the *VIZ Render User Reference.*

20. Save the scene.

21. On the Rendering toolbar, click the Quick Render icon (see Render Stats 06.03).

Render Stats 06.03

Render Time: 2 minutes 06 seconds
Radiosity Processing Time: 1 minute 21 seconds
Music Selection: "In Trutina" Charlotte Church – Voice of an Angel
Color Plate: NA
JPG File: Rendering 06.03.jpg

CEILING

For the ceiling, we will use the same technique as above to create a new material that will simulate a knock-down ceiling.

Create a New Material from Scratch

For this material, we will start from scratch using a supplied template and then configure various parameters to achieve the desired effect.

1. On the Scene - Unused tab of the Materials palette, right-click on Create New tool and choose **New > Paint Flat**.

2. Scroll to the bottom of the list, right-click on the newly created material, and choose **Properties**.

3. Change the name to **MVIZr-Knock-Down-Ceiling**.

4. Change the color to pure White—R:**255**, G:**255**, B:**255**.

5. In the Adjust Radiosity rollout, change the Reflectance Scale to **50** and the Indirect Bump Scale to **10**.

6. In the Special Effects rollout, change the Bump to **25**, and then click the Bump slot.

7. From the Material/Map Browser dialog that appears, choose Noise, and then click OK.

8. Change the name of the Noise map to **MVIZr-Bump-Noise-Knock-Down**.

9. Change the Noise Type to **Fractal**, set the Noise Threshold High to **0.55**, and set Low to **0.3**.

10. Change the Size to **.5** (see Figure 6–13).

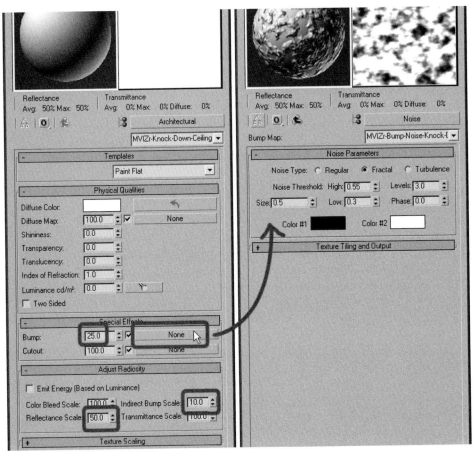

Figure 6–13 *Settings for Ceiling Material*

11. Go to Parent, and in the Texture Scaling rollout, set the Width and Height to **3'-0"**.

Apply the Material to the Scene

12. Drag the new MVIZr-Knock-Down-Ceiling material into the scene and drop it onto the ceiling object.

    When you see a tip indicating Space <Standard> Ceiling, release the mouse.

13. Save the scene.

14. On the Rendering toolbar, click the Quick Render icon (see Render Stats 06.04).

You may notice that the ceiling is not very white. In reality, white surfaces take a lot of their color from reflected light off of adjacent surfaces. This may not *look* like what you would expect, but it is very close to reality.

Render Stats 06.04

Render Time: 1 minute 33 seconds
Radiosity Processing Time: 1 minute 21 seconds
Music Selection: "Tanz Der Schwane" Tchaikovsky – Swan Lake
Color Plate: NA
JPG File: Rendering 06.04.jpg

TRIM

For the trim, we will create a glossy white paint material.

Create a New Material

We will again create a new material from a provided template and then configure it.

1. On the Scene - Unused tab, right-click on Create New tool and then choose **New > Paint Gloss**.

2. Scroll to the bottom of the list, right-click on the new material, and choose **Properties**.

3. Change the name to **MVIZr-White-Trim**.

4. Change the color to pure White—R:**255**, G:**255**, B:**255**.

5. In the Adjust Radiosity rollout, change the Reflectance Scale to **50**.

Apply the Trim Material

Now let's apply this new material to the trim objects. There are several trim objects, so let's open up the selection floater and use it to build a selection and then apply the material.

6. Open the Selection Floater.

7. Place a checkmark in both the Display Subtree and Select Subtree checkboxes.

8. Clear all types in the List Types area, and then select only the Geometry checkbox.

9. Hold down the CTRL key and select the following:

    Both Door <Hinged – Single – 6 Panel>

    Door/Window Assembly <Classic Double Hung + Transom>

    Both Door/Window Assembly <Entry Door + Transom>

10. Click the Select button.

Leave the Selection Floater open.

11. On the Scene - Unused palette, right-click on MVIZr-White-Trim and choose **Apply to Selected**.

12. In the Selection Floater, without the CTRL key, click on the first Wall <Living Room> Baseboard object.

13. Hold down the CTRL key again, and then select all of the remaining Wall <Living Room> Baseboard objects and then click the Select button.

14. Click the Scene - In Use tab on the Materials palette.

 NOTE Since we have now applied the trim material to an object in the scene, it has moved automatically from Scene - Unused to the Scene - In Use palette.

15. Right-click on MVIZr-White-Trim and choose **Apply to Selected**.

The windows are actually built in the ADT model from a Door/Window Assembly. The Assembly defines the three window bays, and nested within each is a separate Window object. We need to apply this material to these nested Windows as well.

16. In the Selection Floater, hold down the CTRL key again, and then click on each of the three Window <Classic Double Hung> objects.

17. Release the CTRL key, and remove the checkbox from the Select Subtree checkbox.

18. Hold down the CTRL key again, and then click each of the three Glass components to remove them from the selection.

19. Click the Select button (see Figure 6–14).

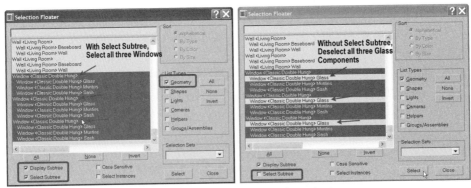

Figure 6–14 *Using the Selection Floater to build a complex selection*

20. Apply the Trim material to the selection and then close the Selection Floater.

21. Save the scene.

22. On the Rendering toolbar, click the Quick Render icon (see Render Stats 06.05).

 Render Stats 06.05

Render Time: 1 minute 34 seconds
Radiosity Processing Time: 1 minute 21 seconds
Music Selection: "Coming Clean" Green Day – Dookie
Color Plate: NA
JPG File: Rendering 06.05.jpg

ENTOURAGE

In this scene, we will place some random, real-life objects—known as entourage—such as having some candles on the table, a beverage in a glass, and something in a bowl. Let's make some materials for these items next.

Create a New Material for Candles

Candle wax is a semi-transparent material. We will start with a semi-gloss paint material and then edit it to have "candle-like" qualities.

1. On the Scene - Unused tab, right-click on Create New tool and then choose **New > Paint Semi-gloss**.

2. Scroll to the bottom of the list, right-click on the new material, and choose **Properties**.

3. Change the name to **MVIZr-Candle**.

4. Change the color to a cream color—R:**245**, G:**234**, B:**190**.

We selected a template to give us some initial settings that would approximate the desired material. Now we need to adjust some of the properties that would better simulate a wax material.

5. In the Physical Qualities rollout, change the Transparency to **10** and change the Translucency to **20**.

6. Using the Selection Floater, select all of the Mass Element <Candles> objects and then apply MVIZr-Candle to the selection.

7. Save the scene.

How about a Beverage?

We have some beverages on the glass table. Let's make them a bit more appetizing.

8. Create a new material based on the Water template and then edit its **Properties**.

9. Change the name to **MVIZr-Beverage**.

10. Change the color to a golden-brown color—R:**220**, G:**215**, B:**90**.

11. Using the Selection Floater, select all of the Mass Element <Liquor> objects and then apply MVIZr-Beverage to the selection.

12. Save the scene.

Something in the Bowl (Flower Petals)

There is also a bowl on the table that needs to be filled with something. Let's try and create a material that looks like flower petals.

13. Create a new material based on the Ideal Diffuse template and then edit its Properties.

14. Change the name to **MVIZr-Flower Petals**.

Let's change the Texture Scaling so we can see the results of our work as we go. We will change it to the appropriate size for rendering when we are happy with what we have created.

15. In the Texture Scaling rollout, set the Width to **100′-0″** and the Height to **100′-0″**.

In many materials (like most of the ones above), we simply pick the diffuse color from the color picker. The diffuse color is the color of the object when directly illuminated. In other words, this is what we would consider the color of the object to be when we look at it. In this case, to achieve the effect that we are after, we will use a Noise map for the diffuse color. This will provide variation in the color, and with the other settings, will allow this material to appear like a bowl full of petals.

16. Click the map slot (currently set to "None") next to the Diffuse Map (see Figure 6–15).

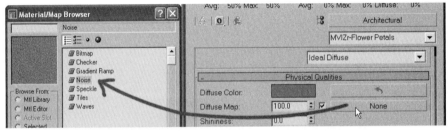

Figure 6–15 *Add a map to the Diffuse Color channel of the new material*

17. In the Material/Map Browser, select Noise from the list and then click OK.

18. Change the name of the Noise map to **MVIZr-Noise-Flower Petals**.

19. Set the Noise Type to **Turbulence**.

20. Set the Size to **1**.

21. For Color #1 set R:**100**, G:**16**, B:**0**.

    This will set a nice, deep red color.

22. Drag the Color #1 swatch and drop it on the Color #2 slot.

23. Click the Copy button in the "Copy or Swap Colors" dialog (see Figure 6–16).

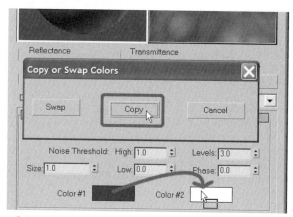

Figure 6–16 *Drag Color #1 to the Color #2 slot to copy it*

24. Click on the Color #2 swatch, and then drag the Whiteness slider down to about the 20% value.

 NOTE Ours was set to R=220, G=190, B=190.

That looks OK as far as a mix of colors go. However, when petals are lying in a bowl you see them from the side, so the colors would be elongated.

25. Click the Go to Parent icon, and then, in the Texture Scaling rollout, set the Height to **10′-0″**.

In the sample at the top of the Material Editor, you can see that the colors are now elongated and approximate what we might see. Since the scaling is too big for our flowers, we need to change it.

26. Set the Width to **10″** and the Height to **1″**.

NOTE We used large values above to make changes evident in the Material Editor slot. These new settings are not likely to show up in the Material swatch. Also, depending on your units settings, feet is likely the default. Therefore, you must use the inch (') symbol when you type in this value.

Even though we will actually "see" the map that we just created as the diffuse color, we still need to set a color for the Diffuse Color slot. We want this to closely approximate the color of the Noise map. The Diffuse Color, and not the Diffuse Map, is what is used by Radiosity and other rendering processing. If we do not choose a matching color, this material will appear to reflect a dull gray color making our rose petals appear like they have been through a failed science experiment. It can be difficult to guess the correct color since the Noise map varies between two colors. Therefore, let's let VIZr do the work for us.

27. Click the left-pointing arrow button directly above the Diffuse Map slot.

This is the texture average button and it will calculate an approximate color within the range of the map.

28. Using the Selection Floater, select the Mass Element <Chips> object, and then apply the MVIZr-Flower Petals material.

29. Save the scene.

30. On the Rendering toolbar, click the Quick Render icon (see Render Stats 06.06).

 Render Stats 06.06

Render Time: 1 minute 35 seconds
Radiosity Processing Time: 1 minute 22 seconds
Music Selection: "The Ocean" U2 – Boy
Color Plate: NA
JPG File: Rendering 06.06.jpg

FLOOR

Many materials have been included with the VIZ Render application. You need only launch the Content Browser to access them. For the floor material, we will search through the Content Browser for an existing carpet material that should work for our scene.

1. Click the Start Content Browser icon on the Tool Palettes toolbar.

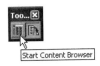

Figure 6–17 *Launch the Content Browser*

2. Click on the Render Material Catalog.

3. Type in **Carpet** in the Search field, and then press ENTER.

 NOTE Should the search yield no results, manually browse the Finishes category, and then the Flooring category.

Let's go with Finishes.Flooring.Carpet.Loop.6 for this scene.

4. Move your cursor over the i-drop icon, press and hold down your mouse button on this icon, and then drag it into your VIZ Render scene.

5. Drop it when you see Space <Standard> Floor on the i-drop tool tip.

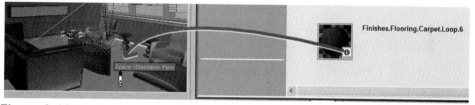

Figure 6–18 *i-drop a carpet material from Content Browser to the scene*

6. Save the scene.

7. On the Rendering toolbar, click the Quick Render icon (see Render Stats 06.07).

 Render Stats 06.07

Render Time: 1 minute 39 seconds
Radiosity Processing Time: 1 minute 26 seconds
Music Selection: "Winter" John Denver – Portrait (Disc 1)
Color Plate: NA
JPG File: Rendering 06.07.jpg

SAVE THE NEW MATERIALS

Since we have spent a lot of time creating new and modifying existing materials, we should store them for future use. The Content Browser is a perfect location for this. As you scroll through the Scene - In Use palette you will see several materials. Some of these came from the original dataset, and some of them were brought into the scene via the furniture that was substituted earlier in this scene.

 NOTE Choosing properties on a material other than an Architectural material that originated in Autodesk VIZ or 3ds max will prompt you to convert the material to a material that can be edited in VIZ Render. Doing so may alter the material so that it renders differently from the original material. The reason is due to the fact that materials in VIZr have a limited number of material editing controls, compared with Autodesk VIZ or 3ds max. If you wish to edit one of these materials, it is recommended that you first copy it and then edit the copy.

1. The Content Browser should still be open. Make it active (click its icon on your Windows task bar, or use CTRL + TAB).

2. Click the Home icon.

3. Right-click in a blank area and choose **Add Catalog**.

4. Choose the Create a new catalog radio button.

5. Change the name to **My MVIZr Materials**. Click the browse button and browse to *C:\MasterVIZr\Catalog* and then click OK twice.

6. Click on the new *My MVIZr Materials* catalog to open it.

7. Click the Create new Category icon.

8. Input the name **MVIZr – Ch06** for the category.

9. Click on the title (which is now a link like a web page) for the new *MVIZr – Ch06* category.

As you can see, this new category is empty. Let's add our Materials to it.

10. From the Scene - In Use palette in the VIZr scene, click and drag the MVIZr-Blinds material and drop it on the Content Browser.

 NOTE The drag-and-drop should be one fluid motion from VIZr to Content Browser. The material will appear within the MVIZr – Ch06 category. If you drop it too soon, you may accidentally drop the material on an object in the scene and thus apply the material to that object. If this happens, choose Undo Set Material from the Edit menu and try again.

 Tip: If you cannot position both VIZ Render and Content Browser side by side to accommodate the drag-and-drop, then, from VIZr, drag to the icon on the Windows task bar, keep the mouse button depressed (which will pop open Content Browser), then move the mouse over the Content Browser window, and finally release the mouse button.

11. Repeat the process for the MVIZr-Stucco Texture material.

12. Using the CTRL key, select each of the remaining materials created in this chapter: MVIZr-Knock-Down-Ceiling, MVIZr-White-Trim, MVIZr-Candle, MVIZr-Beverage, and MVIZr-Flower Petals.

13. Drag any one of the highlighted materials and drop it on the Content Browser.

This will add all of the selected materials to the *MVIZr – Ch06* category within the Content Browser (see Figure 6–19).

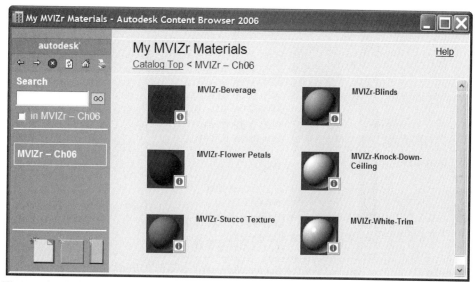

Figure 6–19 *The MVIZr – Ch06 category within Content Browser containing all of this chapter's custom materials*

In this exercise, we added only materials that we actually created in this chapter. As an alternative, you could instead drag the entire Scene - In Use palette to Content Browser. To do this, click and hold down the Scene - In Use tab. Drag it to the *MVIZr – Ch06* category within Content Browser just like you did for the individual materials. The difference will be that the entire palette will be created in the Content Browser. You can then right-click this palette icon in Content Browser, choose Properties, and edit its name and other Properties. To use this palette in VIZr, simply drag it back to your scene. When you do this, the palette will be set to "Refresh" from the Catalog. A small icon, as shown in Figure 6–20, will appear at the lower-right corner of the palette. If the version of the palette in the catalog should change, simply click this icon to retrieve the latest version. This is a very useful way to share materials amongst project team members.

Figure 6–20 *Dragging a Palette from the Content Browser sets it to Refresh from the Catalog*

For more complete coverage of the Content Browser, refer to the online help, or pick up a copy of *Mastering Autodesk Architectural Desktop* by Paul F. Aubin.

RENDERING THE SCENE

Rendering this scene at mid-range and high resolution will take time. There are a lot of reflective materials, specifically the glass. The glass in the scene is supplemental and not the primary focus of the scene. In order to test some of these materials, we had to make some important changes to the Rendering settings in the Preset used above. For quicker renderings, the following settings helped speed up draft renderings and represent a drastic reduction in quite a few settings, as compared to what would be considered production-quality renderings. For more information on any of these settings, refer to Appendix C and the *VIZ Render User Reference*.

MID-RANGE RENDERING

In this next test render, we will take a look at the settings that were contained in the Preset used above and what changes need to be made to bring our rendering to the next level. A preset with the changes suggested here is available in the Render Preset folder, if you prefer.

Explore the Preset

1. On the Rendering menu, choose **Render**, and then click the Common tab.

The Common Parameters rollout includes the frame (or frames) that you wish to render. In Chapter 9 we will explore this in further detail when we explore animation. Beneath that are the options for the rendering size. We have seen these in previous chapters an increase in the size of the rendering generated will naturally have an impact on rendering times. For instance, doubling both the width and height of the output size, will generate four times as many pixels requiring typically four times the render time. We will accept all defaults here for now. In the Render Output rollout, you can choose to render directly to a file or files rather than to the Render Frame Window. For single frame still renderings (not animation) it makes sense to render to the Render Frame Window and then save from there if you are satisfied with the result. You can also choose to do both if you wish.

2. Expand the Photorealistic Rendering rollout.

The **Antialias Geometry** setting was turned off to save render time(as was noted above). From the *VIZ Render User Reference*: "Antialiasing smoothes the jagged edges that occur along the edges of diagonal and curved lines when rendering." This is not important for test renderings. Turning it off above helped speed our processing time.

3. In the Photorealistic Rendering rollout, place a checkmark in the Antialias Geometry checkbox.

For studying the lighting and setting up a scene, the Shadow/Material Quality setting is not as important, therefore it was set to zero to save on rendering time. Let's increase it now to increase the quality.

4. Set the Shadow/Material Quality to **5**.

If the Mapping setting is disabled, you will only see the diffuse color for all the materials. No diffuse, bump, or cutout maps will show up. Since we wanted to see how the stucco material came out, we left this enabled. If you are performing test renders where material display is not critical, you can turn this off.

The **Shadows** setting is especially important for this scene, as we are concerned with testing the daylight streaming in from the windows. Make sure that this is turned on.

5. Expand the Render Options rollout.

Since this scene has a lot of reflective surfaces, this is the most critical render time-saving setting in this type of scene. Of course, with the **Reflections/Refractions** setting disabled, the glass in this scene looked smoky and not much like actual glass. This was not that detrimental for test renderings, but it should now be enabled.

6. Place a checkmark in the Reflections/Refractions checkbox.

All of the scenes in this book use Radiosity. Interior scenes especially benefit from processing Radiosity solutions, because a majority of the light in an interior scene is reflected light. In the **Use Radiosity** preset used above, we adjusted the Radiosity-specific settings to decrease rendering times.

Two-Sided Materials is a global setting that can also be enabled for individual materials. For the final rendering, we will need to turn this one on. You may have noticed when you ran the draft renderings that the arm of the couch was "see through." Apparently there is a "flipped" face in that model. Each face of a 3D Model has a "front" side and a "back" side. When looking at the front, the face appears solid. When looking at the back, it appears invisible. We will need to take this into consideration for the final rendering. It would be preferable to simply enable this setting for this one material only. However, this material has not been converted to a VIZr material and may not render as expected if it were converted. Therefore, to be safe, we will render Two-Sided Materials in the final. This will however add some render processing time.

For a scene with a lot of reflective/refractive materials, adjusting the **Ray Bounces** setting will save a tremendous amount of rendering time. If Reflections/Refractions is disabled, this setting has no affect. If Reflections/Refractions are desired, a Ray Bounces setting of 1 is a good place to start. For the final rendering of this scene, a setting of 6 will be used. For now we will use something a bit lower.

7. Set Ray Bounces to **4** (see Figure 6–21).

Normally **Compute Radiosity when Required** will be enabled. If a setting in the scene has been changed that invalidates the Radiosity solution, it will be recalculated automatically for the next rendering. If this is disabled, this will not occur and you will need to reset and reprocess the solution manually.

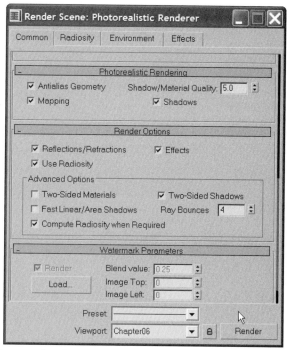

Figure 6–21 *Settings on the Common tab of the Render Scene dialog*

8. Click the Radiosity tab.

9. Expand the Radiosity Processing Parameters rollout.

Radiosity is an iterative process. With each iteration, the solution gets progressively better. You use the **Initial Quality** setting to decide to what level of quality VIZr should process the Radiosity solution before stopping. (Keep in mind that you can never attain 100%.) A good average number is 85%, which will yield very fine results. Adding some Refine Iterations for the final rendering will give quite reasonable results. In the preset above, we had set this to only 60%.

10. Set Initial Quality to **85**.

Refer to the *VIZ Render User Reference* for more information on the additional settings in the Process section.

The **Indirect** and **Direct Light Filtering** setting smoothes the shading variations between each mesh object. Setting this to 3 or 4 is usually sufficient to lessen the triangulated look in the test renderings. We can lower this setting a bit for our mid-range rendering. Later, for the final rendering, we can increase it.

11. In the Interactive Tools area, set both Filtering settings to **1**.

12. Expand the Radiosity Meshing Parameters rollout.

In order to show varying levels of light on larger single faces such as walls, floors, sidewalks, exterior walls, and so on, the Meshing Size will vary depending on the type of scene. Smaller interior scenes can handle smaller meshing sizes, whereas larger exterior scenes use larger meshes. Each object can have its own meshing override, so consider a larger global mesh and smaller mesh setting for specific objects. The default settings here will work fine for this scene.

13. Expand the Rendering parameters rollout (see Figure 6–22).

Sometimes when processing a Radiosity solution, you will end up with light leaks or other visual artifacts where surfaces come together. The light will appear to "bleed" at these seams. The **Regather Indirect Illumination** setting can often greatly clean up these so-called "light leaks" and other artifacts. However, this comes at a great cost in rendering time. If you have the time, the results are well worth the wait. Refer to the *VIZ Render User Reference* for additional information.

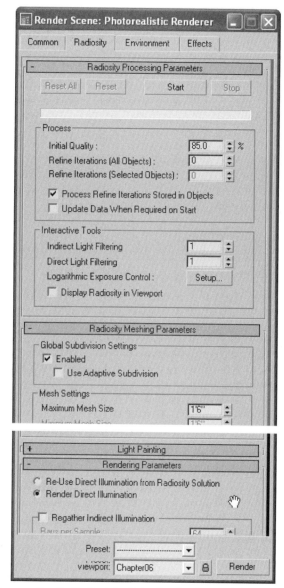

Figure 6–22 *Settings on the Radiosity tab of the Render Scene dialog*

Load the Mid-Range Preset

Now that we have discussed the many settings, let's load a Preset to be sure that all of the settings are configured as outlined here and that nothing was missed.

14. In the Render scene dialog, load the *Ch06_Final Rendering_Mid Range.rps*.

15. In the dialog that appears, be sure that everything is selected, and then click the Load button.

16. In the "Select Preset Categories" dialog, click Yes.

"Advanced Lighting Plugin" is another name for Radiosity. It is a more generic way of describing it. This message is warning you that you will have to reprocess your Radiosity based on the settings in this Preset. This is fine, since you would want to reset your solution before rendering again anyhow.

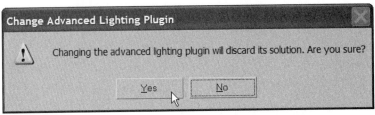

Figure 6–23 *Loading the Preset will invalidate the existing Radiosity solution*

17. On the Rendering toolbar, click the Quick Render icon (see Render Stats 06.08).

If you leave the Render Scene dialog open, you can watch the Radiosity progress on the Radiosity tab. You will see a progress bar with each iteration; as it completes, the solution quality level will update. When the desired quality has been attained, the rendering will process using this solution.

Render Stats 06.08

Render Time: 6 minutes 42 seconds
Radiosity Processing Time: 4 minutes 08 seconds
Music Selection: "Take My Hand" Dido – No Angel
Color Plate: NA
JPG File: Rendering 06.08.jpg

Adding Supplemental Lighting

That took longer than all of test renders, to be sure, but still was not too long. Let's see what it looks like with the can lights and candles turned on.

18. Open the Selection Floater and choose Lights on the right side only.

19. Select **Can EC 01** and **Can Hall 01** and then close the Selection Floater.

20. On the Modify Panel, in the General Parameters rollout, place a checkmark in the Light On checkbox.

NOTE Each light group was created by using the Instance option. So, any change to one of the lights is reflected in the other instances.

21. On the Rendering toolbar, click the Quick Render icon (see Render Stats 06.09).

 ### Render Stats 06.09

Render Time: 2 minutes 54 seconds
Radiosity Processing Time: Reuse Existing Solution
Music Selection: "She's A Lady" Tom Jones — The Best of...Tom Jones
Color Plate: NA
JPG File: Rendering 06.09.jpg

If you look closely, you can see the effect of the can lights on each side of the TV. While you may want that light to be more pronounced, this closely approximates real life. Where you can see it better is if you look under the table behind the couch. You can plainly see the shadow that is created by one of the hallway can lights.

FINAL RENDERING

If you are satisfied with the results of the Mid-Range Rendering, you can skip this section. In this final topic of the chapter, we will load a Preset that will yield even higher quality results. Above we mentioned a few of those settings, such as a re-gather. Feel free to load the Preset and explore the settings on the Common and Radiosity tabs before you actually perform the render. This preset loads a background image (*MVIZr - Ocean Background.jpg*) and turns on the Regather settings. With these settings, your render time should be around 5 to 10 hours.

22. Open the Render Scene dialog and then load the *Ch06_Final Render_High Res.rps* Preset.

23. Click the Render button (see Render Stats 06.10).

 ### Render Stats 06.10

Render Time: 11 hours 56 minutes 32 seconds
Radiosity Processing Time: 12 minutes 17 seconds
Music Selection: Watch "Gone with the Wind (1939)" and then
"Casablanca (Two-Disc Special Edition) (1942)" on DVD
Color Plate: C–6
JPG File: Rendering 06.10.jpg

CONCLUSION AND SUMMARY

In this chapter we have explored the editing and creation of several materials. Using a Noise procedural map, we were able to simulate some interesting materials to apply to the surfaces of our Living Room dataset. We have seen that a collection of materials can be saved in the Content Browser for later retrieval, and we have explored a great many of the Render Scene dialog controls. Among the topics that we covered, we learned the following:

Simple Blocks and Symbols inserted into a scene in ADT can be substituted in VIZr for highly detailed 3D artifacts, complete with material assignments.

Substitution works equally well on 3D and 2D blocks.

A Daylight System can be used to light the interior as well as the exterior.

To use a Daylight System for interior spaces, be sure to choose No in the Daylight Object Creation dialog.

Adjusting the Radiosity settings of existing materials can prevent their reflected color from over-influencing the scene.

You can use the Noise Procedural map to represent all sorts of organic textures, like stucco and knock-down ceilings.

Adding some basic entourage with simple materials can add life and interest to the scene.

Sometimes existing out-of-the-box materials from the Content Browser are perfectly suitable to use as-is in your scene.

Be sure to explore the out-of-the-box offerings in the Content Browser before creating your own materials from scratch.

With careful consideration of each Render setting, you can create simple and quick study renderings and highly detailed and high-quality final renderings.

Settings such as Regather, while capable of adding considerable realism and quality to a rendering, will also require much greater rendering times.

CHAPTER 7

Exterior Nighttime Rendering

INTRODUCTION

In this chapter, we will revisit the townhouse dataset and explore the creation of an exterior nighttime rendering. Using a combination of artificial lighting as well as the Daylight System configured to simulate moonlight, we will create a convincing nighttime scene. This will be the last chapter that uses the townhouse dataset that we constructed in Chapters 3 and 4. If you did not complete those chapters, and wish to understand the various modeling techniques that were used to construct the dataset used in this chapter, please return to Chapters 3 and 4 now and complete them first. However, you are not required to build the model yourself, or even complete Chapters 3 and 4 at all in order to complete this chapter. Feel free to start right here if you are anxious to get right to nighttime rendering.

OBJECTIVES

The primary objective of this chapter is to produce a simple exterior nighttime rendering using various light sources. In this chapter, we will build upon the work we did in Chapter 5, where we rendered the townhouse façade in daylight. Using the Camera angles that were established there, this chapter will focus on the lighting and rendering settings needed for the nighttime effect. In this chapter, we will explore:

- Simulating moonlight with a Daylight System
- Adding exterior artificial lighting using manufacturer's IES files
- Create an exterior nighttime Rendering
- Add a Lighting Effect

THE TOWNHOUSE DATASET

The project for this chapter is a two-story townhouse. The model is provided as a series of Constructs in the ADT Project Navigator. All of the Constructs have been provided, including the façade of the townhouse, which was modeled in Chapters 3 and 4. If you did not complete those chapters, you can visit them now or later to gain a complete understanding of how the model was constructed.

In Chapter 5, we rendered this dataset in daylight. In this chapter, we will use the Cameras and Daylight System (to represent the moon) created there as a starting point. If you completed Chapter 5, you'll recall that we made several Cameras. Feel free to render from any of these, or create your own after the chapter is complete.

Start VIZ Render and Load the Townhouse Project

1. If you have not already done so, install the dataset files located on the Mastering VIZ Render CD ROM.

   Refer to "Files Included on the CD ROM" in the Preface for information on installing the sample files included on the CD.

2. Launch VIZ Render from the icon on your desktop.

   You can also launch VIZr by choosing the appropriate item from your Windows Start menu in the Autodesk group. In Windows XP, look under "All Programs"; in Windows 2000, look under "Programs." Be sure to load VIZ Render, and not Architectural Desktop, for this exercise.

3. From the File menu, choose **Open**.

4. Click to open the folder list and choose *My Computer* and then your C: drive.

5. Double-click on the *MasterVIZr* folder, then the *Chapter07* folder, and finally the *Views* folder.

6. Double-click *Rendering – Front Perspective.drf* to open this scene file.

 NOTE If you wish to explore the model dataset in ADT, please remember that this is an ADT Project. Therefore, you will need to use Project Browser on the File menu to load the **Townhouse07** Project and make it current. If you wish to do this, please refer to the start of Chapters 3, 4, and 5 for detailed steps.

Create a Copy of the VIZr File

The scene that we have opened is the file that we completed at the end of Chapter 5. We will simply save the file as a new name to begin our nighttime scene. In this way we will maintain the link back to the same Project View file in ADT while retaining the changes we made to the VIZr file in Chapter 5 separately. Should the ADT model geometry change, we could thereby update both the daytime and nighttime scenes from the same model file.

7. From the File menu, choose **Save Copy As**.

8. In the Save File As Copy dialog, change the File name to *Townhouse – Nighttime*, and click the Save button.

This will create a new file (with the new name) in the same location, but does not load it. The *Rendering – Front Perspective* file remains active.

9. From the File menu, choose **Townhouse – Nighttime** from the history list.

 NOTE The file appears on the history list, even though we have not actually opened it yet. We do have to open it, however, as "Save Copy As" merely creates the copy. It does not load it.

SETTING-UP NIGHTTIME RENDERING

We will be working with the existing cameras in this file that were created in Chapter 5. In order to simulate nighttime conditions, we will place artificial lighting in several locations and modify the existing Daylight System to simulate moonlight.

Set-up Moonlight

Let's start with the moonlight. Since this file was saved from the existing Chapter 5 file, we can simply modify the Daylight System and edit its parameters to simulate the light from the moon. This is not an exact science. We are shooting for a believable effect here, not necessarily a *scientifically verifiable* solution.

1. Select the Daylight System. (Use the Selection Floater and then select the [Daylight01] object.)

2. On the Modify panel, change the name to **Moonlight**.

Let's adjust the moonlight a bit so it's not in the exact same place as the sun was today (or rather, in Chapter 5).

3. Set the Hour to **12** and the Month to **1**.

 NOTE Again, we are "fudging" this. If we really set the time of day to nighttime, the Daylight System object, which thinks it is the Sun, would drop below the horizon. This would not make a very believable solution.

4. In the IES Sky Parameters rollout, remove the checkmark from the ON checkbox to turn it off.

5. In the Control Parameters rollout, place a checkmark in the **Manual Override** checkbox (see Figure 7–1).

Figure 7–1 *Manually Override the Daylight Settings to configure it for Moonlight*

In general terms, sunlight is about 275 times brighter than moonlight. To simulate this, we need to reduce the intensity considerably. Now that we have switched to a Manual Override, we can simulate this by reducing the Intensity setting.

6. In the Sun Parameters rollout, set the Intensity to **50**.

We should also change the color of the light to simulate the bluish hue that moonlight generates.

7. In the Sun Parameters rollout, click on the color swatch to the right of Intensity.

8. In the Color Selector dialog, set Red: **195**, Green: **200**, Blue: **246**, and then click Close (see Figure 7–2).

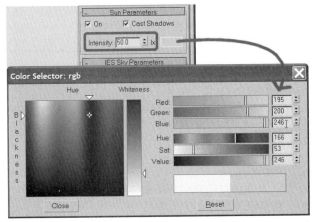

Figure 7–2 *Change the intensity and the color of the Daylight object to match Moonlight*

Perform a Test Render

We are just about ready to perform a render to see the effects of our moonlight settings. Since this will be a test rendering, we should dial back some of the rendering settings saved in the file from Chapter 5.

9. Open the **Render Scene** dialog.

10. On the Common tab, in the Photorealistic Rendering rollout, remove the checkmark from the Antialias Geometry checkbox.

11. In the Render Options rollout, remove the checkmark from the Reflections/ Refractions checkbox (see Figure 7–3).

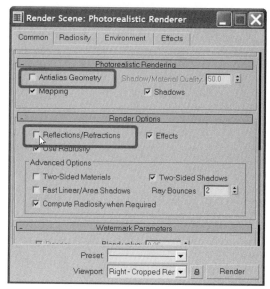

Figure 7–3 *Change a few render settings to make the test rendering process more quickly*

12. Verify that the Refine Iterations (All Objects) is set to **0** (see Figure 7–4).

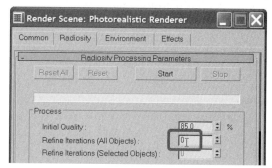

Figure 7–4 *Reset the Radiosity solution and adjust the parameters*

13. On the Environment tab, click the Color swatch for the Background.

Let's choose a very dark blue color for the background to help create a convincing nighttime scene.

14. Set the color to Red: **10**, Green: **10**, Blue **50**, and then click the Close button.

This will give us a nice deep blue sky more appropriate for nighttime.

15. In the Logarithmic Exposure Control Parameters rollout, set the Brightness to **70**, and uncheck Exterior daylight (see Figure 7–5).

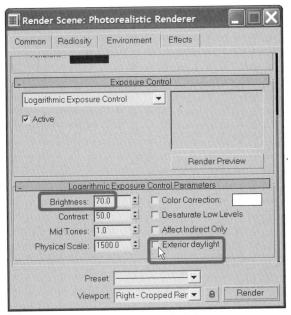

Figure 7–5 *Adjust Brightness and turn off Exterior daylight*

We are adjusting the brightness a bit to help bring out the details in the scene. Otherwise the scene will be too dark. Remember the iris analogy? Our iris must open to allow more light in. We have turned off the Exterior Daylight option because this is now moonlight, and we do not wish to have it calculate in the same way as it would for daylight.

16. Click the Render button, and then click the OK button at the bottom-right corner of the viewport to accept the indicated cropped region (shown in Figure 5–35 in Chapter 5). See Render Stats 07.01.

Render Stats 07.01

Render Time: 11 minutes 6 seconds
Radiosity Processing Time: 10 minutes 40 seconds
Music Selection: "All Blues" Miles Davis – Kind of Blue (Remastered) (1997)
Color Plate: NA
JPG File: Rendering 07.01.jpg

This scene looks pretty good so far. The moonlight is fairly convincing, but the scene is still rather dark. In the next topic, we will begin adding some additional artificial lights to help spruce things up a bit.

ADDING LIGHTS

Unless the inhabitants of our townhouse are out for the evening, there would probably be some lights turned on in the interior. In addition, there are carriage lights at the front door on the exterior, and the street lamps that we worked with in Chapter 5.

Adding Interior Lights and Window Sheers

To add some lighting inside the townhouse, let's merge in some lights from another file. Also, to give our homeowners some privacy, we will also add some curtains to the windows. Adding curtains will also save us the need to add furniture to the inside of the space, which adds a lot of faces that need to be rendered and that slow down rendering times. The sheers will mask the fact that the inside of the townhouse is empty! A single merge file contains both the lights and curtains. The curtains were created from a simple extruded spline.

1. Switch the current viewport to the **Center – Camera Correction** camera view.

2. From the File menu, choose **Merge**.

3. Browse to the *C:\MasterVIZr\Chapter07\Elements* folder, and then double-click the *Curtain Merge.drf* file to open it.

4. In the File Merge dialog, click the All button (at the bottom-left of the dialog) and then click OK (see Figure 7–6).

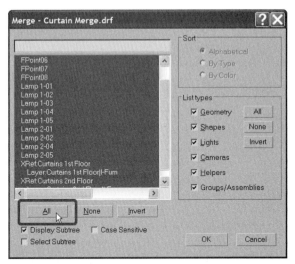

Figure 7–6 *Click the All button to merge all objects in the Curtain Merge.drf file*

Peek at each viewport and notice that several objects have appeared inside the townhouse. For more information on the Merge feature, please refer to Chapter 8.

5. On the Rendering toolbar, change the Render Type from Crop to **View** (the opposite is shown in Figure 5–34 in Chapter 5).

6. Click the Render Scene icon, and change the Height for the Output size to **800** and the Width to **600**.

 NOTE You may need to click the aspect lock icon to disable it. Otherwise the Width will change proportionately.

7. Click the Render button (see Render Stats 07.02).

 NOTE You should not need to click OK this time. If the Crop boundary appears again, cancel the rendering and then repeat Step 5.

 Render Stats 07.02

Render Time: 12 minutes 21 seconds
Radiosity Processing Time: 11 minutes 21 seconds
Music Selection: "Piano Concerto In D Minor, K.466: 1st Movement" Neville Marriner
– Amadeus Original Soundtrack Vol 2.
Color Plate: NA
JPG File: Rendering 07.02.jpg

Carriage Lights

Let's illuminate the two carriage lights on either side of the front door. We can install a simple 100-watt light bulb in each one. This is easy to do using the preset 100-watt bulb objects. This task will be a bit easier from a Top viewport.

> If you have a viewport maximized, on the keyboard, press ALT + W to minimize it.

8. In the Top viewport, zoom in on the left carriage light at the front door.

 TIP If you want to make the top viewport less "busy," toggle off the Radiosity meshing. To do this, choose RADIOSITY from the Render menu, and then deselect the Display Radiosity in Viewport checkbox in the Radiosity Processing Parameters rollout.

9. From the Create menu, choose **Photometric Lights > Preset Lights > 100W bulb**.

10. Place the light in the center of the carriage light (see Figure 7–7).

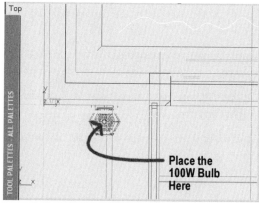

Place the
100W Bulb
Here

Figure 7–7 *Place a Preset 100W Bulb light in the carriage light*

11. In the Modify panel, change the name to **Carriage Left**.

12. From the Tools menu, choose **Align**. Align the light to center it (X and Y) on the carriage light.

 NOTE See the "Align" topic in Chapter 2 for more information on the Align Tool.

13. In the Left viewport, move the light up to fit inside the bulb of the carriage light (see Figure 7–8).

NOTE Our light ended up at X=0′-11 1/2″,Y= -0′-3″, Z=12′-1″.

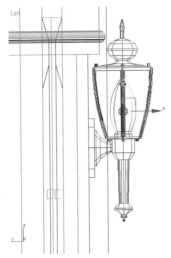

Figure 7–8 *Align the light to the X and Y center of the carriage light and then move it up to the bulb's location*

14. In the top viewport, clone an instance of the light for the right-side carriage light. (Hold down the SHIFT key while you drag the X-axis to the right.)

15. In the Clone Options dialog, make sure Instance is selected, change the name to **Carriage Right**, and then click OK.

16. Zoom in as required, and use the align tool again to center it in the right carriage light (see Figure 7–9).

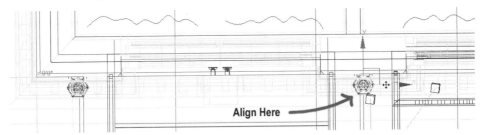

Align Here

Figure 7–9 *Create an Instance of the carriage light, and then align it to the carriage light on the right*

The height should be fine since you copied it in the XY plane. More information on cloning and align can be found in Chapter 2.

Lamppost Lights

For the streetlights, we will download the IES file from a manufacturer's website so that we have the most accurate light distribution information. An IES file simulates the actual "real life" light pattern of the light fixture.

17. In the top viewport, zoom in on the leftmost lamppost.

18. From the Create menu, choose **Photometric Lights > Free Point**.

19. Click to place the Free Point light in the scene next to the lamppost.

The lamppost has two lamps; one over the street and the other over the sidewalk.

20. Relative to the lamp over the sidewalk, use the Align tool again and align the light Center to Center in both the X- and Y-axes, and then click the Apply button (see Figure 7–10).

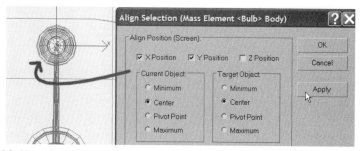

Figure 7–10 *Align the light to the center of the sidewalk lamp relative to Top view*

21. Remaining in the Align Selection dialog, choose Z Position and align the Center of the Current Object to the Minimum of the Target Object (see Figure 7–11).

 TIP Consult the Left viewport to verify that the position is aligned correctly.

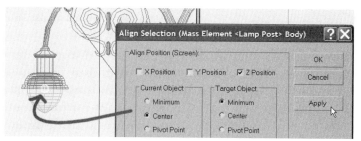

Figure 7–11 *Align the light to the Target Minimum in the Z-direction*

In the Left viewport, Move the light down slightly.

NOTE Our light ended up at Z=13'-7".

22. In the Modify panel, change the name to **Lamppost Left 01**.

23. Switch back to the Top viewport (be sure to right-click for this so that the light stays selected), and then from the Tools menu, choose **Array**.

24. In the Incremental area, in the Move row, change the X value to **29'-0"**.

Be sure that Type of Object is set to Instance (the default).

25. In the Array Dimensions area, change the 1D Count to **3** and the 2D Count to **2**.

26. For Incremental Row Offsets, input **–5'-0"** in the Y field and then click OK (see Figure 7–12).

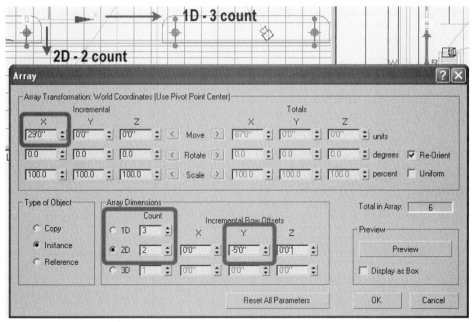

Figure 7–12 *Use the Array tool to create six instances of the light*

The Incremental value is used with the transform selected—Move, in this case, to space each item. Arrays can create objects in one, two, or three dimensions in the same operation. Since we have three lampposts, with two lamps each, we need a total of six instances, or a three-by-two array. Since we set the Incremental value for X, the 1D Count applies to X—in this case, three—and the Y applies to the 2D Count, or two, in this case.

NOTE The two lights at the right may need to be moved a bit to make them align with the lamppost on the right.

Now that we have our lights positioned, we can load an IES file to set a realistic distribution of light in this situation. We can go out to the World Wide Web, visit a lighting manufacturer's web site, such as Kim Lighting or Lightolier, and find a real-life light that we could use for this scene. We have included such an IES file on the CD ROM in the *Maps* folder.

Be sure that one of the lights is still selected from the previous sequence.

27. On the Modify Panel, in the Intensity/Color/Distribution rollout, choose **Web** from the Distribution list.

28. In the Web Parameters rollout, click the <None> button to the right of Web File.

29. Browse to the *C:\MasterVIZr\Maps* folder, choose the *Sr4f517m.ies* file, and then click Open.

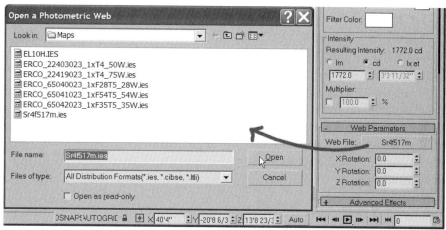

Figure 7–13 *Load a manufacturer's IES Photometric web file*

Notice that all six lampposts changed, since they are all instances.

Adjust Rendering Settings

Let's load another preset to change some render settings. Feel free to study the dialog to see what has changed. The preset we are going to use increases the material quality and ray bounces slightly; it also adjusts the Radiosity settings to increase the refines and decrease the filtering. Rendering with this preset will take a bit longer, so you may want to go out for movie.

30. In the Render Scene dialog, load the *Ch07_Final Render_Mid Range.rps* preset.

Be sure all items are selected when you click Load, and answer Yes to confirm the Advanced Lighting warning.

31. Save the scene.

32. From the Render Scene dialog, click the Render button (see Render Stats 07.03).

Render Stats 07.03

Render Time: 34 minutes 28 seconds
Radiosity Processing Time: 15 minutes 09 seconds
Music Selection: Gipsy Kings – Gipsy Kings (1988) Entire CD
Color Plate: NA
JPG File: Rendering 07.03.jpg

CAUTION *Please don't close the Render Frame window yet. We will need the rendering open for the next sequence.*

ADDING AN EFFECT

You may notice in some night shots that the lights seem to glow. In the upcoming steps we will apply an *Effect* in VIZr to simulate this real-world occurrence. The effects update the current image in the Rendered Frame Window. If you have not completed a render or have closed your Rendered Frame window, your scene will be rendered again with the current settings when you apply the effect. You may want to load a draft preset and render if you don't want to wait for that final image when initiating the effect from the following steps.

Add a Blur Effect

1. From the Rendering menu, choose **Effects**.

2. In the Effects rollout, click the Add button.

3. In the Add Effect dialog, choose Blur (see Figure 7–14).

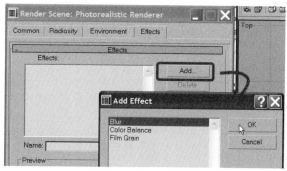

Figure 7–14 *Add a Blur Effect*

4. Click the Update Effect button.

You will see the progress bar move its way down the render frame window again. At first it will appear as if nothing is occurring. The Blur is applied to the rendering after the complete frame is processed. To see the blur progress interactively, you must place a checkmark in the "Interactive" box. When the Effect update is complete, notice that the entire image has been blurred. That is not at all what we are after.

5. Place a checkmark in the Interactive checkbox (see Figure 7–15).

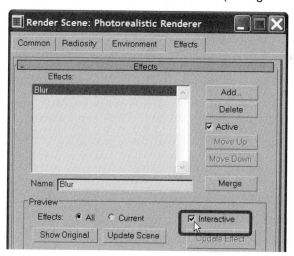

Figure 7–15 *Turn on the Interactive Feature to see the effect as you make changes*

6. Accept the defaults on the Blur Type tab, and then click the Pixel Selections tab.

7. Remove the checkmark from the Whole Image checkbox.

As we mentioned above, we don't want the whole image blurred, just the areas around the lights. Since the lights are indeed the brightest objects in our scene, let's apply the blur only to the brightest objects.

8. Place a checkmark in the Luminance checkbox (see Figure 7–16).

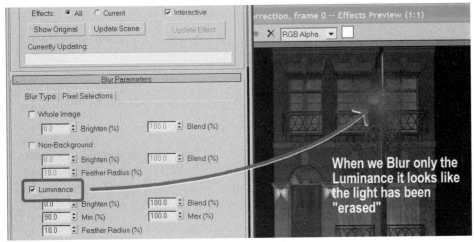

Figure 7–16 *Set the Blur to affect only the Luminance*

We are now seeing some blurring around the lights, but it looks as if we took an eraser to our rendering.

9. Change Brighten (%) to **450**, and then press ENTER.

OK, this is a step in the right direction, but now it is much too extreme. Let's soften the Effect a bit. By increasing the Brightness percentage, we have brightened the pixels that we are blurring. In this way, it doesn't simply smear the nearby pixels creating the muddy effect we had above, but rather, considerably brightens the area as it blurs it.

10. Change Blend (%) to **50**, and then press ENTER.

That softens the Effect by blending it in with the nearby pixels. That is much better, but it could use one more tweak.

11. Change the Feather Radius (%) to **7**.

This will lessen the distance that the blur travels from the bright objects. The Feather Radius is basically how far out to nearby pixels the Effect can reach. Feathering is a kind of blending. By reducing the radius of this blending, we make the Effect slightly more contained to its immediate area.

TIP Try setting the Feather Radius to about 60. This will make the scene appear like a foggy night.

Perfect—almost. Notice how the bright spots on the lamppost are softening as well. That's not exactly what we had in mind. We need to have this effect only apply to the lights themselves. We can do this via the material applied to the glass of the light fixture (see Figure 7–17).

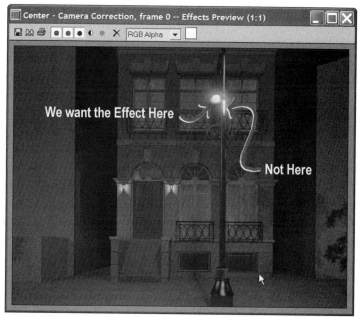

Figure 7–17 *The Effect is almost perfect, except that it applies to all bright spots, not just the light*

Using Material Effects Channels

We can make all of the previous settings apply on a by-material basis. To do this, we use Material Effects Channels.

Leave the Render Scene dialog and the Current Render Frame window open.

12. On the Scene – In Use palette, right-click the MVIZr – Street Globe and choose **Properties**.

13. Just beneath the Reflectance values at the top of the Material Editor, click and hold the Material Effects Channel flyout (currently set to 0).

14. Roll your cursor over the **1** slot and then release the mouse (see Figure 7–18).

Figure 7–18 *Assign the Material to Special Effects Channel 1*

This action makes this material unique to all the other materials in the scene. If we apply our Blur Effect only to Channel Number 1, this will be the only material in our scene to be affected.

Adjust the Effect

Now let's make the Effect apply to this Material Channel instead.

15. Return to the Render Scene dialog.

16. Remove the checkmark from Luminance.

17. Scroll all the way to the bottom of the Pixel Selections rollout.

18. Place a checkmark in the Material ID checkbox, type 1 in the ID field, and then click the Add button.

The values on the right are a global setting for this Blur effect. If you would like varying settings for other material IDs in the scene, simply create another Blur effect. You can create a unique name for each in the Name section at the top of the Effects tab.

19. Use the same values from the Luminance Blur above to the Material ID section for ID 1 (see Figure 7–19).

▶ Change Min Lum (%) to: **50**.

▶ Change Max Lum (%) to: **100**.

▶ Change Brighten (%) to: **450**.

▶ Change Blend (%) to: **50**.

▶ Change F. Radius (%) to: **7**.

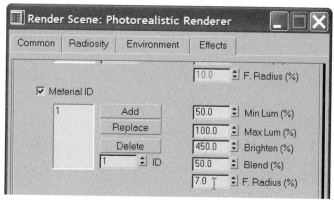

Figure 7–19 *Transpose settings to the Material ID section*

20. To see the effect, render the scene again (see Render Stats 07.04).

Render Stats 07.04

Render Time: 34 minutes 28 seconds
Radiosity Processing Time: 15 minutes 09 seconds
Music Selection: Fleetwood Mac – Rumours (1977) Entire CD
Color Plate: NA
JPG File: Rendering 07.04.jpg

FINAL HIGH-RESOLUTION RENDERING

Assuming that you are happy with the results you have gotten up to now, you may want to generate a high resolution rendering of this scene. To do this, we have provided a preset. This preset modifies some of the options on the Radiosity tab, specifically, the Regather settings. Rendering with this preset will take quite a long time. You may want to let it run overnight. For more information on Regather settings, refer to Chapter 6 and the online Help.

High-Res Option 1

1. Open the Render Scene dialog and load the *Ch07_Final Render_High Res Render Preset*.

   As before, load all options and answer yes to confirm resetting the Radiosity (Advanced Lighting) solution.

2. Save the scene.

3. Click the Render button (see Render Stats 07.05).

TIP This rendering took nearly 9 hours for us. You will want to let it run at a time when you can be away from your computer, for instance, overnight is usually best.

Render Stats 07.05

Render Time: 8 hours 25 minutes 48 seconds
Radiosity Processing Time: 1 hour 12 minutes 32 seconds
*Music Selection: Watch the original "Star Wars Trilogy
(Widescreen Edition)" on DVD (2004) (or watch the prequel trilogy when "Revenge of
the Sith" comes out on DVD...)*
Color Plate: C–8
JPG File: Rendering 07.05.jpg

High-Res Option 2

If you would like to take that final rendering to the next level, consider making the following changes in the scene as an additional exercise.

4. Select the Moonlight and change the Rays per Sample setting to **30**.

If you recall, we changed the Rendering Properties for the adjacent buildings, sidewalk, and street in Chapter 5. Because we decided not to mesh those objects, the light levels across those surfaces may not vary as they would in real life. If you like, you can turn meshing for them back on now.

5. Select the buildings to the left and right as well as the sidewalk, curb, and street.

TIP It may be easiest to do this in Top view.

6. Right-click and choose **Rendering Properties**.

7. Put a checkmark next to the Use Global Subdivision Settings box and the Subdivide box beneath that.

8. Modify the Meshing Size to be **5′-0″**.

This will increase the time required to process the Radiosity solution but will not affect the actual time to render the scene.

9. Save the scene.

10. Open the Render Scene dialog and Render the scene.

Plan for about 8 to 10 hours total rendering time.

CONCLUSION AND SUMMARY

In this chapter we built on the dataset that we have been refining over the last several chapters. Revisiting this now-familiar dataset, we gave it a very different look by changing the time of day and experimenting with some effects. In this chapter we have learned the following:

The Daylight System can be adjusted to simulate Moonlight.

When simulating Moonlight, you must enable manual override and "fudge" the time of day settings so that the Daylight System does not dip below the horizon.

Adding some simple elements to the interior, such as lights and curtains, can make the rendering more realistic.

Place photometric lights within the exterior light fixtures to supplement the moonlight.

Use a Manufacturer's IES file to simulate accurate light distribution patterns.

Render times increase dramatically when enabling the high-quality Radiosity settings such as Regather, but the results are worth it.

Plan your long renders to run overnight.

Effects such as a Blur can be added to the bright lights in the scene to add realism.

CHAPTER 8

Interior Space with Artificial Light

INTRODUCTION

In this chapter, we will explore the creation of an interior rendering using artificial lighting. The exercise covered here is a nighttime scene, but you can use the techniques covered here for both daytime and nighttime rendering. The image generated in this chapter is the one we used for the cover of this book. In this chapter we will use fixtures specifically chosen from the out-of-the-box offerings for this scene, as well as some additional entourage to generate another great version of this rendering.

OBJECTIVES

In nearly every chapter of this book, we have used the Daylight System as the primary light source. The main objective of this chapter is to discuss how to achieve excellent results without the use of the Daylight System Object. In this chapter we will explore:

- Working in interior spaces
- Placing and Configuring Artificial Lighting
- Placing i-drop Content in a scene
- Setting up a Background Image

THE CONFERENCE ROOM DATASET

The Conference Room model is a simple ADT model composed of Walls on three sides and a Curtain Wall on the fourth. A Space object is used for the floor and the ceiling. Within the room is a conference table made from a few Mass Elements. Mass Elements have also been used for a drywall ceiling feature and the can lights scattered throughout the space.

Start VIZ Render and Open the Conference Room Scene

1. If you have not already done so, install the dataset files located on the Mastering VIZ Render CD ROM.

Refer to "Files Included on the CD ROM" in the Preface for information on installing the sample files included on the CD.

2. Launch Autodesk Architectural Desktop from the icon on your desktop.

   You can also launch ADT by choosing the appropriate item from your Windows Start menu in the Autodesk group. In Windows XP, look under "All Programs"; in Windows 2000, look under "Programs."

 NOTE Be sure to load Architectural Desktop, and not VIZ Render, for this step.

3. From the File menu, choose **Open**.

4. Click to open the "Look In" folder list, and choose *My Computer* and then your C: drive.

5. Double-click on the *MasterVIZr* folder, and then the *Chapter08* folder.

6. Double-click *Conference Room.dwg* to open this file (see Figure 8–1).

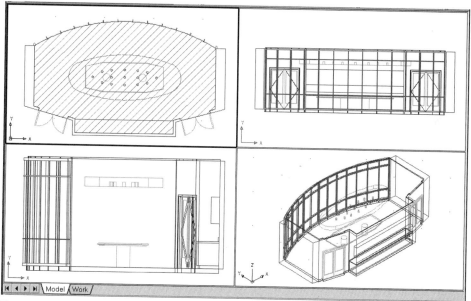

Figure 8–1 *The Conference Room file has been saved with four viewports showing it from several views*

Take a few moments to explore the makeup of this file; just be certain not to edit any objects. If you change anything, be sure to undo before you proceed. You can click on the various objects and then look at their properties in the Properties Palette. This is an early schematic design proposal that was created quickly in ADT. Many of the objects do not yet even have materials assigned.

Explore the Dataset

7. Select one of the Walls, right-click, and choose **Edit Wall Style**.

8. Click on the Materials tab (see Figure 8–2).

   Notice that all of the components have only the "Standard" material assigned.

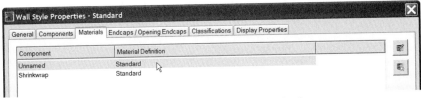

Figure 8–2 *Nearly all Materials assigned in this file are Standard*

Repeat this process on other objects if you wish. Most objects in the drawing have only the Standard material. This is typical of what we might have in the early schematic design phase of a project. The designer would be more concerned with the overall layout and form of a space, and may not have given much thought yet to the actual Wall Styles, Construction, or Materials that would ultimately be required.

Despite these apparent limitations in the model, it is actually very easy to link this model to VIZr and quickly apply materials. We have seen examples of this already in Chapter 6. More importantly, as we saw in Chapter 2, we can decide later upon future reloads whether to keep the VIZr scene's material assignments, or to replace them with the ones being imported from the ADT Model link. In this way, we make it very easy to start using VIZ Render very early in the design process without concern that we might lose carefully crafted materials in VIZ Render when the ADT model changes.

> If you are satisfied that you have a good understanding of the ADT model, we can move on to VIZ Render.

 NOTE If you have made any changes to the ADT model, be sure to Undo them before proceeding.

9. From the Open Drawing Menu, choose **Link to VIZ Render** (see Figure 1–2 in Chapter 1).

The Link to VIZ Render has been previously created, and, therefore, when we link to it now, the existing *Conference Room.drf* file will open and the File Link Settings dialog will appear. This behavior was discussed previously in the "Working With File Link" topic of Chapter 2.

10. In the File Link Settings dialog, click the OK button (see Figure 8–3).

The scene should appear with viewports minimized. If it did not, minimize them now so you can see all four views.

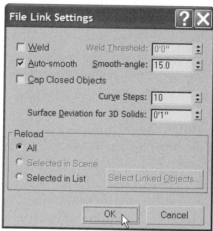

Figure 8–3 *The File Link Settings dialog will appear, confirming the reload of the ADT model into the VIZr scene*

It looks like one of our colleagues has already begun working on this scene. Notice that the scene includes two additional items that were not included in the original DWG file. These are a roll of drawings sitting atop the conference table and a Camera object. The Camera has been added for convenience. If you wish to add your own Cameras, or would like more information on Camera creation and manipulation, please refer to Chapter 5.

 NOTE The materials used on the roll of drawings have bitmaps assigned to the various slots. Bitmap paths used by the Mastering VIZ Render datasets were discussed in the "How to Use The CD" topic of the Preface. If you did not follow the instructions there to configure a path to the *C:\MasterVIZr\Maps* folder, you will receive an error dialog about missing bitmaps when this dataset loads. To fix this problem, read and follow the instructions in the Preface to add the required path.

The roll of drawings was created in a separate ADT model from AEC Polygons and Mass Elements. A DRF was created from this model and a bitmap material of the drawings applied to the top sheet. This DRF was then merged into the current scene and positioned on top of the conference table using the align command.

We will be adding several additional items to this scene using the same process below in the "Adding Entourage" topic.

SETTING-UP THE SCENE

As we have mentioned in the introduction, the goal of this exercise is to create a nighttime rendering using only artificial lighting. If you look at the Top viewport, you will notice that Camera will be looking toward the window wall. For a more interesting and dramatic rendering, we will use a photograph of the Chicago skyline for the backdrop. This photo will be seen through the windows beyond the rendering. Let's start by looking through our Camera, and then we will add the skyline image.

Set-up the Camera

1. Right-click in the Perspective viewport to make it active.

2. On the keyboard, press the **C** key.

   This should make the **Skyline Camera** active in place of the Perspective view.

With the exception of the image that appears on the surface of the roll of drawings, the space appears rather gray and dreary. In the coming steps we will address this. For the moment, let's adjust the Camera view. The lens is set a bit narrowly, where a wide angle would be more appropriate for this interior space and show us more of the scene.

3. In the Top viewport (in the lower-left corner of the room), click to select the Camera object.

4. On the Modify Panel, in the Parameters rollout, click the 24mm Stock Lens button (see Figure 8–4).

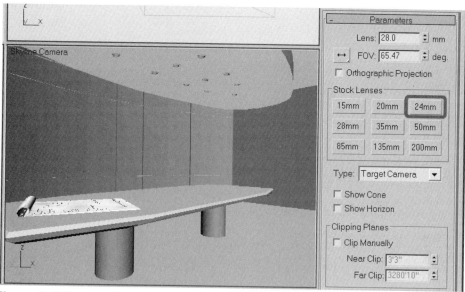

Figure 8–4 *Choose a 24mm Stock Lens for the Camera View*

The scene is a bit "gray." Naturally we will need to apply some materials to this scene to brighten things up a bit. In the next sequence, we will set up the skyline background image. However, in their present state, we won't be seeing much out of these windows. Even though we will add materials to the scene in the topics that follow below, let's apply a glass material to the windows now so that we can see outside.

5. Click the "Doors & Windows" palette tab to make it active.

NOTE If you do not see this tab, it may be hidden. Look for the small group of bunched up tabs at the bottom of the palette and click there to locate the Doors Windows palette. (see Figure 2–4 and Color Plate C–2.)

6. Locate and drag the Doors & Windows.Glazing.Glass.Clear tool and drop it on top of one of the window panes (see Figure 8–5).

When you see a tip appear reading "Curtain Wall <Conference Curtain Wall> Default Infill," release the mouse.

NOTE If a message appears indicating that the material already exists, choose **Replace it** and then click OK. This dialog appears because the material is assigned to the double door behind the camera already. If you prefer, you can avoid this message by dragging the material from the Scene – In Use palette instead.

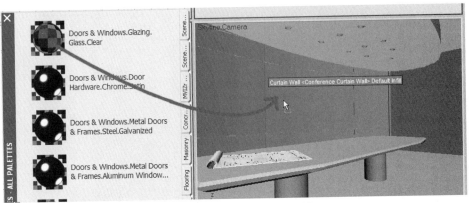

Figure 8–5 *Drag the Clear Glass material to the Window Infills*

The windows should now be transparent.

    7. Save the scene.

Set-up the Background

Outside the window wall, we want to see the city skyline. In this case, we have a photograph of the Chicago skyline to display here in the distance. We could use an Environment map for this (via the background control in the rendering dialog), but we have found that doing so has some limitations. It turns out that the portion of the image that should be reflected (such as within the highly reflective tabletop), does not actually reflect the appropriate section of the image. Instead, the final image will appear as if the background is bleeding through from behind the reflective sections. (This limitation also exists in Autodesk VIZ and 3ds max®.) You can see this problem illustrated in the left side of Color Plate C–15a.

Therefore, in order to work around this problem, we will merge in a DRF file that contains a large Wall object. When viewed from the room, this wall would appear as a large movie screen off in the distance. We will use the surface of this wall to map a photograph of the cityscape in the distance beyond our conference room's windows. As a result, there will actually be an object in the scene and the rendering's reflections of this object will appear correctly. You can see this problem illustrated in the right side of Color Plate C–15a.

    8. From the File menu, choose **Merge**.

    9. Browse to the *C:\MasterVIZr\Chapter08* folder, and then double-click the *BackgroundPlane.drf* file.

    10. In the Merge dialog, click the All button, and then click OK (see Figure 8–6).

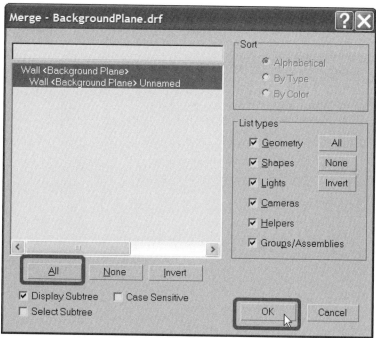

Figure 8–6 *Merge all objects from the BackgroundPlane.drf file into the current scene*

You should see a large plane appear outside the window. This wall has been sized quite large (100′ x 65′), so that it can be pushed far away from the scene. This is done so that lights from the inside will not reflect on the wall's surface. To be sure that it does not adversely affect our rendering time, we will adjust a few more settings.

11. Open the Selection Floater, highlight the "Wall <Background Plane> Unnamed" object beneath the "Wall <Background Plane>" object, and then choose Select.

    Close the Selection Floater.

12. On the Modify panel, rename this object to **Background Plane**.

13. With this object still selected, right-click, and choose **Rendering Properties**.

14. In the "Geometric Object Rendering Properties" area, remove the checkmark from the Cast Shadows checkbox.

15. In the "Geometric Object Radiosity Properties" area, remove the checkmark from the Cast Shadows, Diffuse, Use Global Subdivision Settings, Use Adaptive Subdivision, and then the Subdivide checkboxes.

16. Place a checkmark in the "Exclude from Regathering" checkbox (see Figure 8–7).

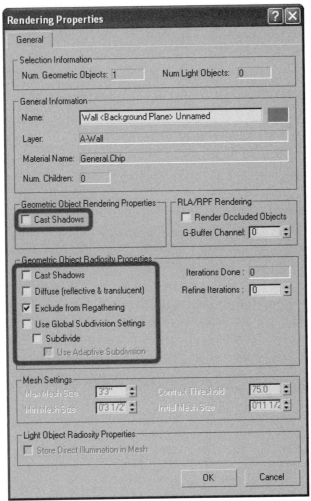

Figure 8–7 *Configure the Rendering Parameters of the Background Plane*

17. Click OK to dismiss the Rendering Properties dialog and then Save the scene.

Creating and Applying the Skyline Material

18. On the Scene - Unused tool palette, create a new material based on the Paper template.

19. Right-click the new material and choose **Properties**. Rename the material to **MVIZr-Skyline**.

20. Click the Diffuse Map slot (currently set to "None") to open the Material/Map Browser.

21. Double-click the Bitmap item and then browse to the *C:\MasterVIZr\Maps* folder. Then double-click the *Chicago-Night.jpg* bitmap file.

22. Rename the Map **MVIZr-Chicago-Night**, and then directly beneath the sample sphere, click the "Go to Parent" icon.

23. Click the "Set color to texture average" button (the small arrow above the map slot).

This control queries the bitmap and determines the average color from the bitmap image and assigns those RGB values to the color. While not strictly necessary for this background material, it is not a bad habit to get into.

24. Change the Luminance to **250**.

You should see the sample sphere brighten slightly. The Luminance setting determines whether a material will appear to glow or be "self illuminating." This can be used for any material that you wish to appear lit, such as neon signs. In this case, we are using this so that the skyline will appear to glow.

25. In the Texture Scaling rollout, right-click on the Width spinner to set the value to **0**. Repeat on the Height spinner (see Figure 8–8).

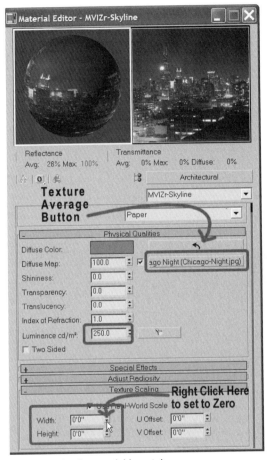

Figure 8–8 *Configure the new Background Material*

Close the Material Editor.

26. Select the Background Plane object and apply the new Skyline material to it.

Adjust the Mapping Coordinates

When you apply the material, you can clearly see that something went wrong. It turns out that the bitmap is not properly mapped to the background plane. We have not yet run into this issue, since as we have seen throughout this book, nearly all ADT objects import to VIZ Render with materials and *mapping coordinates* automatically applied. Mapping Coordinates are used to align and size bitmap images along the surfaces of 3D geometry. Normally a Wall, such as our background plane here, will know what to do on its own. However, since this is a *very* large wall, and we are basically mapping the entire cityscape upon it, we will need to adjust its default mapping. To remedy the situation, we will apply a UVW mapping modifier to the wall plane. To make it easier to see what is happening as we work, let's change the view in the Left viewport.

27. Right-click in the Left viewport to set it active, and then click the Default Southwest View icon on the View Orientation toolbar.

    Further Adjust the Zoom as Necessary to see the entire scene comfortably.

28. Right-click the viewport label (now labeled "User"), and choose **Smooth + Highlight**. Zoom Extents in the viewport (see the top-left side of Figure 8–9).

 NOTE See the "Viewports" and the "Viewport Navigation" topics in Chapter 2 for more information on these steps.

29. With the Background Plane still selected, click the UVW Map button on the Modify panel (see the top-right side of Figure 8–9).

That did not seem to help very much. However, once we adjust a few parameters, we should be all set.

A UVW Modifier is basically a set of instructions that tell the bitmap how to apply itself to the selected geometry. Several options are available. We will use Planar, since we are looking directly at a single plane of the wall and are only concerned with the way the bitmap is applied to that plane of the wall.

30. On the Modify panel, with the Background Plane still selected, scroll down to the Parameters rollout and make sure that Mapping is set to Planar.

31. Change the Length to **64'-5"** and the Width to **100'-0"**.

    This is still not quite right. Since we are using Planar mapping, we need to adjust the alignment.

32. In the Alignment area, choose the Y radio button (see the bottom-right corner of Figure 8–9).

33. In both the Camera and User viewports, right-click the viewport label and choose **Texture Correction**.

These sizes are the width and height of the wall that we merged into this scene. We want the bitmap image to fit this plane as precisely as possible so that it does not repeat the image. When you create a bitmap to use in a material, such as bricks or concrete, you typically want it to repeat or "tile" across the surface of the model. In this case, we only want to see the Chicago skyline once. When we shot the photograph, we simply measured the width and height of the photograph and built the wall in the merge file to match the same proportions. You can be a little "off," but you want the proportions to be as close as possible so that you don't have to stretch the photograph to keep it from repeating. The reason that we used the Y Alignment is simply because Planar mapping gives us only the Length and Width parameters, but we wanted our texture to be applied vertically. By changing the alignment you are basically rotating the orientation to the desired plane.

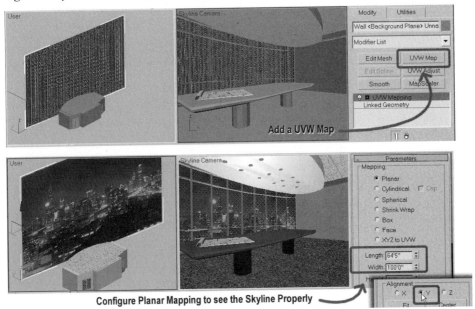

Figure 8–9 *You can now see the Chicago Skyline beyond, through the windows*

    34. Save the scene.

ADDING MATERIALS

At this point we have added a very interesting background image to the scene, but we still have quite a plain room. The next thing that we will do is add some materials to the scene. As you can see, most of the objects in the scene have the General Chip material applied. We need to finish the Curtain Wall by applying a material to its mullions, as well as add the carpet and paint materials. We will also create some materials for the wall paint and trim, as well as for the Conference table and base.

Curtain Wall Mullions and Frames

    1. Open the Selection floater and select all of the Curtain Wall <Conference Curtain Wall> components, except for the Default Infill (see Figure 8–10).

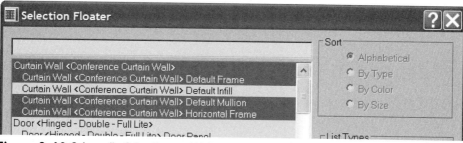

Figure 8–10 *Select all of the Curtain Wall Mullions and Frames*

2. From the Doors Windows palette, right-click the Doors & Windows.Door Hardware.Chrome.Satin and choose **Apply To Selected** (see Figure 8–11).

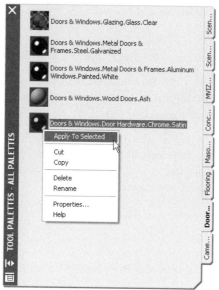

Figure 8–11 *Apply an existing material to the Curtain Wall components*

Floor and Ceiling

Let's take a trip to the Content Browser for the Floor and Ceiling Materials.

3. Click the Start Content Browser icon on the Rendering toolbar (see Figure 6–17 in Chapter 6).

4. Click on the Render Material Catalog.

5. Type in **Carpet** in the Search field, and then click Go.

 NOTE Should the search yield no results, manually browse the Finishes category, then the Flooring category.

6. Using the i-drop icon, drag the Finishes.Flooring.Carpet.Floral material from the Content Browser and apply it to the Floor of the space.

7. In the Content Browser, click the Back icon, then type **Ceiling** in the Search field, and then click Go (see Figure 8–12).

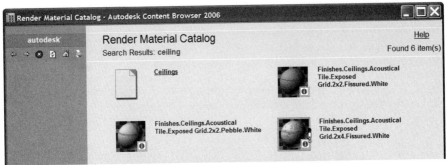

Figure 8–12 *i-drop a Ceiling material to the Space object's Ceiling Component*

8. Using the i-drop icon, drag the Finishes.Ceilings.Acoustical Tile.Exposed Grid.2x4.Fissured.White material from the Content Browser, and apply it to the Ceiling of the Space object.

 NOTE Be sure that you are dropping the material on the Space object's ceiling and not the Mass Element ceiling feature in the center of the room. The tool tip should read Standard <Ceiling> before you release the mouse.

Walls

Let's paint the Walls.

9. Create a new material from the Paint Semi-gloss template and Name it **MVIZr-Walls**.

10. Set the Diffuse Color to pure white: R:**255**, G:**255**, B:**255**.

11. Apply the material to the soffit—Mass Element <Cloud> Body.

12. Using the Selection Floater, select all of the Wall <Standard> objects, and then apply the white semi-gloss paint material.

13. Create another new material from the Paint Gloss template and Name it **MVIZr-Wall Base**.

14. Set the Diffuse Color to pure white: R:**255**, G:**255**, B:**255**.

15. Apply the new material to all of the baseboard objects—Wall <Base> *and* Wall <Base> Unnamed.

NOTE If you need more detailed steps to create these Materials, review the "Materials" topic in Chapter 6.

Conference Table

The room is beginning to look better. All we need now is a material applied to the conference table.

16. Create a new material from the Wood Varnished template.

17. Edit the Properties and rename the material **MVIZr-Conference Table**.

18. Change the Diffuse Color to a dark brown: R:**80**, G:**60**, B:**20**, and then apply it to the Conference Table—Mass Element <Conference Table Top>.

19. From the Paint Gloss template, create a pure black (R:**0**, G:**0**, B:**0**) material named **MVIZr-Conference Table Base**.

20. Apply it to the Conference Room Table Bases—Mass Element <Conference Table Base> Body (see Figure 8–13).

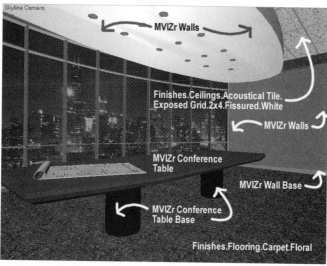

Figure 8–13 *Finish assigning materials to the objects in the scene*

21. Save the scene.

ADDING ENTOURAGE

We really ought to have some chairs to go with our conference table. As we said in Chapter 2, Herman Miller has a nice selection of i-drop content. Let's return to their web site to import a conference chair.

Adding the Conference Room Chairs

1. Using your Web Browser, go to *www.hermanmiller.com*.

2. From the "Choose a Site" list, click *Architects & Designers*. From the Browse list on that page, click *Planning and Visualization*, and then click *3D Models*.

3. From the 3D Models list, choose **Multipurpose Chairs** (see Figure 8–14).

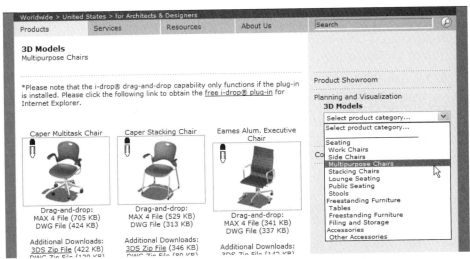

Figure 8–14 *Access conference room chairs from the Herman Miller web site*

Herman Miller Web Site and i-drop Content used with permission from Herman Miller, Inc.

4. i-drop the Eames Softpad Executive chair into your scene.

If you do not have a live Internet connection, you can access instead a small collection of artifacts used in this book from the Content Browser. Included with the files installed from the Mastering VIZ Render CD ROM are all of the furniture and other items referenced in the exercises in this book. To access these local versions, open your Content Browser and add a new catalog. Change the file type to HTM, and then browse to the *C:\MasterVIZr\Catalog* folder. Load *Index.htm*. Browse this catalog after you load it, and then click the Herman Miller category. Locate the Eames Softpad Executive chair and then i-drop it into your scene.

5. When prompted, choose **Merge** from the small menu and then click anywhere to place the chair.

6. Move and Rotate the chair to a desirable location.

7. Clone additional copies (12 total) and place them around the table (see Figure 8–15).

 TIP Use the SHIFT key while moving to clone one or more copies. You may also use the Array and mirror tools on the Tools menu. For more information on cloning, refer to the "Tools Menu" topic in Chapter 2, and the online help.

 TIP For additional realism, move the chairs around a bit after placing them, and rotate some of them a few degrees to make it look like people have sat in the chairs recently.

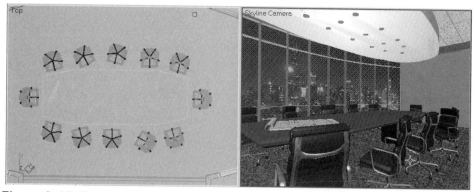

Figure 8–15 *Clone and position 12 chairs around all sides of the conference table*

8. Save the scene.

Place Items on the Table

There is a study model to go along with the roll of drawings on the table. These items are being placed in the conference room in preparation for the big client meeting. The study model is a simple massing model constructed from Mass Element objects in a separate drawing file. It is not possible to link more than one drawing to the same DRF file in VIZ Render. However, as we have seen, we can merge another DRF file into the current one. There is no limit to the number of file merges you can perform on a single DRF, so what we will need to do is open the drawing file in ADT first, link it to VIZr to create a DRF, and then merge this DRF back into the conference room.

Leave VIZ Render running for this exercise.

9. If you have closed ADT, launch it now and open the file named *Study Model.dwg*.

This small model was created in ADT using Mass Elements. To save time in this chapter, we have provided the ADT model for you.

10. From the Open Drawing Menu, choose **Link to VIZ Render**.

   This will launch a second session of VIZr and load the ADT model there.

11. From the File menu, choose **Save**, and then close this session of VIZr.

With these simple steps, we now have some geometry that we can merge into other VIZr scenes. Any time you would like to add entourage to your VIZr scenes, you can search the web for MAX and DRF files to i-drop as we did with the chairs, or you can build the item yourself in ADT, link it to its own scene in VIZr, and save it to create a new DRF. This DRF (and any DRF) can be merged into any other VIZr scene.

   Return to the Conference Room session of VIZ Render.

12. From the File menu, choose **Merge**.

13. In the Merge dialog, click the All button to select all incoming objects, and then click OK.

The study model will appear in the scene but it is not positioned properly.

   The study model object should still be selected.

14. In the Top viewport, use Move to position it nearer the center of the table.

15. From the Tools menu, choose the **Align** tool.

16. At the Pick Align Target Object prompt (in the Status Bar), click the table top in any viewport.

17. In the Align Selection dialog, place a checkmark in the Z Position checkbox to align the object along the Z-axis.

18. Choose Minimum for the "Current Object" (the study model in this case) and choose Maximum for the "Target Object" (the conference table in this case), and then click the Apply button (see Figure 8–16).

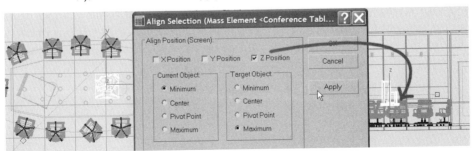

Figure 8–16 *Use the Align Tool to place the study model on top of the table*

19. Click OK to dismiss the Align Selection dialog.

20. With the study model still selected, right-click and choose **Rotate** from the Edit/Transform Quad menu.

21. Rotate the study model to about **-80°** along the Z-axis.

 NOTE There is nothing special about the -80° rotation. The authors simply find this angle pleasing. Feel free to rotate it to another angle if you wish.

As we have mentioned, you can merge items from any DRF file into any other. In Chapter 6, we had lots of small items placed on the various tables within the living room scene. Many of those would add a nice touch to this scene. The only difference in the process from what we have performed so far is that we choose *only* those objects that we wish to merge, rather than clicking the All button.

22. Choose **Merge** from the File menu.

23. Browse to *C:\MasterVIZr\Chapter06* and select the file that you saved at the end of that chapter.

24. Using your SHIFT and CTRL keys, select only those objects that you wish to merge from this file.

For instance, you may want to choose the Mass Element <Tray>, or the Mass Element <Beverage>, or even Mass Element <Glasses>. You can merge as many of these objects into your scene as you wish. Use the same techniques covered above to move, rotate, and align them on top of the table top. If you have other DRF files on your system, feel free to borrow objects from them as well.

In some cases, when you merge in an item from another file, a Duplicate Name dialog will appear. This happens when the names of the items in the file you are merging from are the same as an item(s) in the current file. You can decide how you would like VIZ Render to process the duplicate names as shown in Figure 8–17.

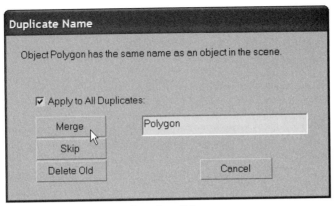

Figure 8–17 *The Duplicate Name confirmation dialog*

The **Merge** button will merge the new object with the new name if you choose to rename it (in the provided text field). It is good modeling practice to create new styles in ADT for different types of objects. In the Living Room scene we have Mass Element styles for Glasses and Beverage. However, there are six glasses and two beverage mass elements. Upon a merge it would be difficult to discern which glass or beverage mass element is being listed. One option is to create a new style for each instance of the object, such as Glass01, Glass02, BeverageGlass, BeverageDecanter, etc. However, that may become quite time consuming, and require more coordination with little benefit. It may be better to choose a unique name after the merge. However, be aware that upon a Reload, that name will revert to the name in the ADT drawing file.

The **Skip** button will not merge the new object.

The **Delete Old** button will merge the new object and will delete the object already in the scene that has the same name. This last option may not be desirable at all. The workflow for VIZ Render is that all modeling is done in ADT and then linked to VIZr. If the original object with the duplicate name is in the scene as part of a file link (as opposed to from an earlier Merge) and is deleted in this step, it will reappear upon a Reload.

Wall Hangings

On the opposite side of the room, the far wall is a bit too bare. Let's add some artwork to that wall. Some wall hangings have been provided in a separate DRF file. Visit *www.paulaubin.com* for a downloadable PDF tutorial on how these were created.

25. From the File menu, choose **Merge**.

26. Browse to the *C:\MasterVIZr\Chapter08* folder and choose the *WallArt.drf* file.

27. In the Merge dialog, click the All button and then click OK.

If the Duplicate Names dialog should appear, place a checkmark in the "Apply to All Duplicates" checkbox, and then click the Merge button to continue.

 NOTE See the passage above regarding Duplicate Names for more information on this.

As you can see, the Wall Art was created to merge directly into the correct location relative to the conference room, so no moving, rotating, or aligning will be necessary. You can do the same in your own scenes by simply building the items that you wish to merge relative to the desired final location in the ADT model. For instance, you could build the items (such as these picture frames) directly in the original ADT model (or a saved-as copy), and then cut and paste them out to a separate drawing file. Next, follow the process above for the study model and link them to a new DRF file. Once complete, you are ready to merge them into your scene and they will maintain the proper location relative to the scene. Of course, you are always free to fine-tune an object's position at any time by simply selecting the items and using any of the VIZ Render transforms to relocate or reorient them.

BUILDING AN OFFICE STANDARD CONTENT LIBRARY

To take the fullest advantage of the Merge feature, you could begin to amass a library of components and 3D artifacts. These items could be in a variety of formats (DWG, DRF, and MAX). However, you may want to develop a consistent strategy for such items. The following factors should be taken into consideration:

- Develop a consistent and logical naming scheme. Names should be easy to identify and descriptive.

- Consider converting all items to a consistent file format. While this is not required, it will make the development of a consistent Merge Procedure easier to establish, repeat, and document.

- Consider building your own internal i-drop pages for the delivery of Content Library items. i-drop accommodates the importation of both the model and all of its associated materials.

- Be sure that all team members have their support paths configured the same. Store all common maps in a common location that is on this search path, so that there is no chance that missing bitmaps will occur.

While the temptation is often great to create a library and then lock it down to all but read-only access for general everyday users, this approach can sometimes create bottle-necks. An ADT and VIZr library will grow all the time as new content becomes available. Develop a consistent process for adding items to the library and encourage all VIZr users to contribute items.

28. When you are finished merging in additional artifacts, Save the scene.

CHOOSING LIGHTS FOR THE SCENE

This is the only project in this book that will not have a daylight system. This scene will use a variety of lights—indirect lights above and below the upper cabinets, indirect lights in the soffit above the conference room, and direct recessed lights above the conference table and wall washers for the wall art on the back wall. We will use some out-of-the-box lights, as well as some i-drop lights from ERCO.

Using Preset Lights

To add some general lighting, we will add some out-of-the-box preset lights for above and below the cabinets on the back wall.

1. From the Create menu, choose **Photometric Lights > Preset Lights > 4ft Cove Fluorescent (web)**.

2. Add the light toward the bottom edge of the Top viewport, next to the built-in cabinets at the bottom wall.

3. On the Modify Panel, rename the light **Lower Cabinet Fill 01**.

4. Clone two more lights (Instances) as shown in Figure 8–18 (be sure that these are named **Lower Cabinet Fill 02** and **Lower Cabinet Fill 03** respectively).

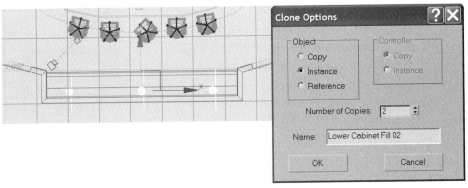

Figure 8–18 *Add three lights at the back wall near the cabinets*

If you do not have the Transforms toolbar loaded, load it now.

5. On the Transforms toolbar, choose **Use Pivot Point Center** from the Pivot Point flyout icon.

6. Choose the **Rotate** transform icon (or right-click and choose **Rotate** from the Edit/Transform Quad menu).

7. Rotate the lights **90°** (see Figure 8–19).

 TIP You can rotate the lights with the mouse and the angle snaps. Right-click the Angle Snap icon and set the Angle (deg) to a value such as 15°. Then be sure that this toggle icon is depressed (pushed in), and your rotations will snap to 15° increments onscreen.

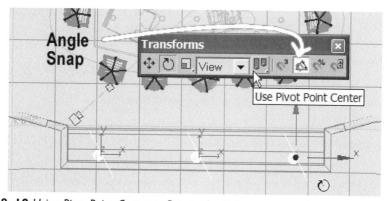

Figure 8–19 *Using Pivot Point Center to Rotate the Lights around their own centers*

If you look in either the Left or Front viewports, you will notice that the lights are on the floor. We can switch to one of those views and move them up.

8. In the Left viewport, switch to the Move transform and move the lights just below the upper cabinet (see the left side of Figure 8–20).

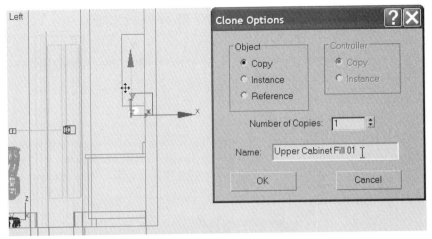

Figure 8–20 *Move the lights to the proper height*

9. Begin moving them up again, hold down the SHIFT key as you move (clone) the three lights up to just above the upper cabinet.

10. In the Clone Options dialog, choose Copy, set the Name to **Upper Cabinet Fill 01**, and then click OK (see the right side of Figure 8-20).

 NOTE Make sure that the other two copies are named Upper Cabinet Fill 02 and Upper Cabinet Fill 03, respectively.

11. Rotate the lights above the cabinet so that they are pointing up to light the wall above the cabinet.

12. Save the scene.

Soffit Fill Lighting

Some other fill lights will be used above the drywall soffit to wash the ceiling above. These will be concealed from the view of the camera but the lighting effect will be evident.

13. From the Create menu, choose **Photometric Lights > Preset Lights > 4ft Cove Fluorescent (web)**.

14. Position the light above the oval soffit in the center of the conference room.

15. On the Modify Panel, rename the light **Soffit Fill 01**.

16. Using Figure 8–21 as a guide, make seven instances (eight total) positioned around the soffit.

TIP Try adding one light in each orientation and then using the Mirror tool on the Tools menu to clone the others. Refer to Chapter 2 for more information.

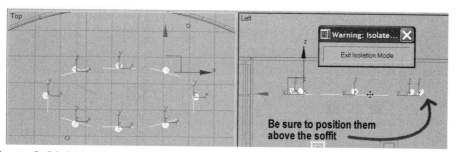

Figure 8–21 *Add eight more lights above the soffit in the center of the room*

Figure 8–21 shows the space without the chairs displayed for clarity. If you wish to work this way as well, select all of the chairs, then right-click and choose **Isolate Unselected** from the Display Quad menu. This will hide the selected items (the chairs in this case). A warning box appears with a large button, shown in Figure 8–21.Click the Exit Isolation Mode button to turn off isolation and restore the chairs when you are ready.

Lighting Data from a Manufacturer's Web Site

If you look closely at the soffit lights we placed above, the portion that represents the light has a specific shape to it. This is because this particular light is associated with an IES file. IES stands for Illuminating Engineering Society. From ERCO's site: "The IES data format is an internationally accepted data format used for describing the light distribution of luminaries. It can be used in numerous lighting design, calculation, and simulation programs." The next series of lights we add will be standard spotlights. We will modify them to use an IES file which we have acquired from the manufacturer's web site.

 NOTE As an alternative to the next several steps, you can access the IES files mentioned from the files installed from the Mastering VIZ Render CD ROM. The required IES files have been included in the *C:\MasterVIZr\Maps* folder.

1. Using your Web Browser, browse to *www.erco.com*.

2. Click the *Products* link, and then choose **by item number > USA, English** from the dropdown list at the bottom.

3. Type in **22403.023**, and then click Show Product.

   We will associate this file with the spots above the conference table.

4. Once you have chosen the model number and loaded the page, click the **Download** link at the top, and then on the left, click the **IES Data link** (see Figure 8–22).

At the top-right you will see a compressed file named *erco_ies_22403023.zip*.

Figure 8–22 *Erco Web Site gives access to hundreds of IES data files and lighting fixtures*

Erco Web Site and IES Content used with permission from Erco, Inc.

5. Click on the compressed file to download it.

6. Save the file to your *C:/MasterVIZr/Maps* folder (If a message appears asking you to confirm overwriting an existing file, choose Yes).

7. On the same page, change the product number to **22419.023**, and then click Show Product.

We will associate this file with the spots for the opposite wall in the next topic.

8. Decompress any ZIP files in the *Maps* folder.

 NOTE If you do not have access to the Internet, or if you encountered any problems with accessing these items directly, all IES files have been included in the C:\MasterVIZr\Maps *folder.*

Adding the Spotlights

Now that we have downloaded the IES file, let's add some spotlights and then associate the IES file with those lights. As an alternative in your own scenes, you can also download the entire fixture from the manufacturer's site if you prefer. Here we already modeled some simple objects to represent the can lights, therefore we simply need to add some standard VIZr lights and then associate the downloaded IES file with those lights in the scene.

9. Maximize the Top viewport.

10. From the Create menu, choose **Photometric Lights > Preset Lights > Halogen Spotlight**.

11. Change the Name of the Light to **Conference Table Spot 01**.

12. Create instances in each of the can lights above the conference room (15 total).

13. Repeat the process to add a new light at one of the smaller spots nearest the far wall named **Accent Spot 01**, and then create other instances (four total) of it for the others (see Figure 8–23).

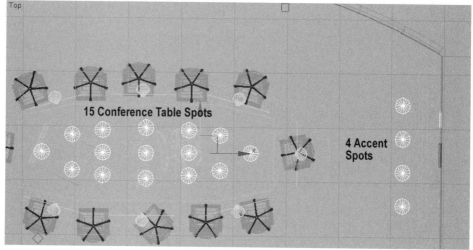

Figure 8–23 *Make two groups of Spotlights, one group of 15 above the Conference Table and the other group of four as Accent lighting to the right*

 Tip: You can Clone Move (hold the **SHIFT**key as you move) the first one from a Conference Table Spot, and use the Copy option in the Clone Options dialog. Then Clone this one, switch to Instance, and set the number of copies to three in the subsequent Clone Options dialog.

Be sure to make Instances of each grouping of lights. This way, when we associate the IES file to one of the conference table spots, all of them will change. Likewise, when we apply the other IES file to one of the wall wash spots, the other three will change.

14. Move the lights up as necessary so that they are just inside (vertically) of the cutout in the soffit (see left side of Figure 8-24).

Associating the IES file

15. Select one of the Conference Table Spot spotlights.

16. In the Modify panel, under the Intensity/Color/Distribution rollout, change the Distribution to **Web** (see right side of Figure 8–24).

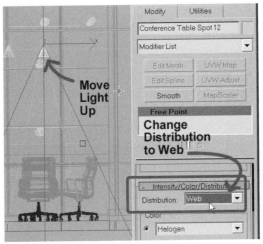

Figure 8–24 *Change the Distribution type of the Conference Table Spots to Web*

 NOTE Notice that the change affects all 15 Conference Table Spots since they are instances. If it does not, then you might need to repeat the clone steps above. The change should *not* affect the four Accent Lights.

17. Beneath the Web Parameters rollout, click the None button to the right of Web File.

18. From your *C:/MasterVIZr/Maps* folder, choose *ERCO_22403023_1xT4_50W.ies*, and then click Open (see Figure 8–25).

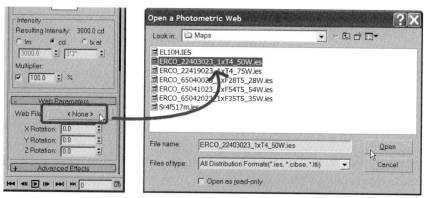

Figure 8–25 *Load a Manufacturer's IES file to set the Web Distribution*

This action sets the 15 spotlights to behave like Erco 22403.023 Light Fixtures. The IES file simulates the true light distribution of the real-life fixture.

19. Repeat the process for the four Accent Spots. This time choose the *ERCO_22419023_1xT4_75W.ies* Photometric Web file.

20. Save the scene.

RENDER THE SCENE

We have applied several materials and created several lights. This is our first scene that does not use a Daylight System. You are probably anxious to see the results. As usual, let's run a test render first. This will allow us to fine-tune anything before waiting for the final rendering to process.

Test Render

We'll use an out-of-the-box Preset for this one.

1. Switch to the Camera viewport if it is not already active.

2. On the Render toolbar, click the Render Scene icon.

3. At the bottom of the dialog, load the Draft with Radiosity Out-of-the-Box Preset to test our scene.

4. In the Select Preset Categories dialog, be sure everything is selected and then click Load.

5. In the Render Scene dialog, click the Render button (see Render Stats 08.01).

Render Stats 08.01

Render Time: 4 minutes 47 seconds
Radiosity Processing Time: 3 minutes 8 seconds
Music Selection: "Family Snapshot" Peter Gabriel – Shaking the Tree: Sixteen Golden Greats (1990)
Color Plate: NA
JPG File: Rendering 08.01.jpg

Mid-Range and High Resolution Renderings

The following Mid-Range Preset adjusts the final rendering proportions to make it a bit wider. It also adjusts some of the Render Options such as ray bounces to make the glass look a little better. Meshing was enabled under the Radiosity tab to better approximate the light in the room.

To Render with this Preset, do the following:

6. Open the Render Scene dialog and then load the *Ch08_Final Render_Mid Range.rps* Preset.

7. Render the Scene (see Render Stats 08.02).

Render Stats 08.02

Render Time: 22 minutes 0 seconds
Radiosity Processing Time: 18 minutes 58 seconds
Music Selection: "Clocks" Coldplay – A Rush of Blood to the Head (2002)
Color Plate: NA
JPG File: Rendering 08.02.jpg

The High Res Preset has further adjusted the ray bounces and meshing parameters to increase the quality of the final rendering. Regather Indirect Illumination was also enabled to help reduce artifacts and shadow leaks. As we mentioned in previous chapters, the use of Regather provides a better quality rendering but will take much longer to render.

To Render with this Preset, do the following:

8. Open the Render Scene dialog, and then load the *Ch08_Final Render_High Res.rps* Preset.

9. Render the Scene (see Render Stats 08.03).

Render Stats 08.03

Render Time: 9 hours 13 minutes 50 seconds
Radiosity Processing Time: 25 minutes 21 seconds
Music Selection: Watch The Lord of the Rings - The Fellowship of the Ring (Platinum Series Special Extended Edition) (2001) on DVD
Color Plate: C–10
JPG File: Rendering 08.03.jpg

CONCLUSION AND SUMMARY

As you can see from this chapter, there are many similarities in the setup process between a scene lit with daylight and a scene lit with artificial light. It takes a bit more effort to light the scene well when using artificial light. Using manufacturer's IES files does help attain realism, but you must carefully plan the location of your light fixtures and test your scene with test renderings as you work. In this chapter, we learned the following:

The Background Environment Map does not work well with reflective surfaces in the foreground. Use a plane and map the background to it to achieve better results.

To create a background backdrop material, you need a photograph and a plane on which to map it. A wall works well for this.

If your background is very large, you will likely need to override the default mapping coordinates using a UVW Map modifier.

Even a very schematic scene can be made quite impressive with a few simple materials.

Adding other entourage, such as glasses, picture frames, and chairs, helps to bring realism to the scene.

Place items at skewed orientations to make the scene more believable.

You can light your scene with a combination of provided photometric lights and lighting data available from manufacturer's web sites.

Using manufacturer's IES data files allows you to bring very realistic light distribution to even out-of-the-box fixtures.

Animation and Client Presentation

INTRODUCTION

In this chapter, we return to the retreat pavilion in which we began our exploration of VIZ Render in Chapter 1. In this chapter we will explore animation and other presentation formats. First we will animate a camera at a given moment in time, and then we will animate the sun and view the effect from a single vantage point over time. Additionally, we will explore other presentation techniques, such as orthographic renderings and panoramic renderings.

OBJECTIVES

The focus of this chapter is animation and other exciting forms of client presentation. In this chapter, we will explore the following:

- Camera animation
- Working with ADT's Live Section
- Working with the VIZr Ram Player
- Panorama Rendering
- Daylight animation

THE RETREAT DATASET

As you may recall from Chapter 1, the Retreat project is a simple model. It is a "Box on the Landscape." The model was built using simple ADT objects. If you would like to see the detailed description of the model, refer to the beginning of Chapter 1. Let's assume that it is now time to present our Retreat Pavilion to our client and we would like to present the design to them in a format other than static "snap-shot" renderings. This will be the premise of this chapter.

START VIZ RENDER AND LOAD THE RETREAT DATASET

1. If you have not already done so, install the dataset files located on the Mastering VIZ Render CD ROM.

    Refer to "Files Included on the CD ROM" in the Preface for information on installing the sample files included on the CD.

2. Launch VIZ Render from the icon on your desktop.

   You can also launch VIZr by choosing the appropriate item from your Windows Start menu in the Autodesk group. In Windows XP, look under "All Programs"; in Windows 2000, look under "Programs." Be sure to load VIZ Render, and not Architectural Desktop, for this exercise.

3. From the File menu, choose **Open**.

4. Click to open the folder list, and choose *My Computer* and then your C: drive.

5. Double-click on the *MasterVIZr* folder, and then the *Chapter09* folder.

6. Double-click *Retreat.drf* to open it.

Producing Variations

Since we may be modifying this file for different purposes as we go, we need to make copies of the file. Doing so will make it easy to explore different options easily. Each copy will have a different name, but they will all be linked back to the original ADT model. This way, if there are any design changes before we complete our presentation renderings, we can simply reload the ADT model and continue working. This is an important concept that is critical to maintaining a dynamic link to the ADT model and the rest of the project team as you explore design and presentation options.

7. From the File menu, choose **Save Copy As**.

8. For the File name type **Retreat – Animation Study**, and then click Save.

9. From the File menu, choose **Retreat – Animation Study** from the recent file list.

   When prompted to save changes to the current file, click No.

USING RPC CONTENT

The first animation we will complete is one where the camera will circumnavigate the model. This type of animation will be much more interesting if the scenery changes as we move around the model. The company ArchVision is a software developer that creates "Rich Photorealistic Content" (RPC) for Autodesk VIZ, 3ds max, and VIZ Render. They produce a specific type of RPC that suits this need. The Environment RPC contains a matrix of Sky, Horizon, and Ground Plane. A watermarked version of this RPC has been provided in the *C:\MasterVIZr\Maps* folder.

INSTALLING RPC

The RPC Plug-in is available as a free download from ArchVision. We have provided the latest version of their plug-in that was available at press time. You can find it in the *C:\MasterVIZr\Plugins* folder. To be sure that you have the latest version, please visit *www.archvision.com*. From the home page, choose Products, and then RPC Plug-ins (see Figure 9–1).

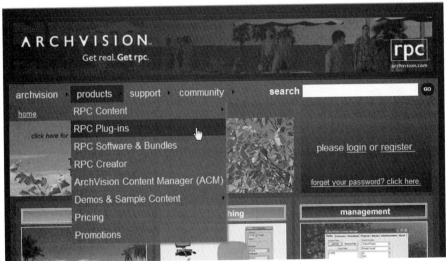

Figure 9–1 *Visit the ArchVision web site for the latest version of the RPC Plug-in*

ArchVision Web Site and RPC Content used with permission from ArchVision, Inc.

From the Page that appears, choose RPC for Autodesk Architectural Desktop, and then download the appropriate plug-in for your version of ADT/VIZr. Please install this plug-in on your system before continuing with the exercise.

When you start the insertion of an RPC it looks for valid RPCs in the same folders that are listed in the VIZr search paths. We have added the *MasterVIZr\Maps* folder to our search path in previous chapters, but let's verify that now.

Verify/Set the Search Path

1. From the Customize menu, choose **Configure Paths**.

The Configure External File Paths dialog will appear. In this dialog is listed all of the currently configured search paths for VIZ Render. Using the buttons on the right, we can modify, delete, add, or change the search order of these paths. When a file is loaded that contains external file references (such as bitmaps), VIZr will search in each of these folders in the order that they are listed in an attempt to locate the required file(s). It is important to note that if different versions of the same file are located in more than one of the listed folders, VIZr will use the first one that it finds. This means that in addition to being certain to add any additional paths not already listed, you may also want to give some consideration to their order within the list.

2. Click on the External Files tab.

In the Preface, in the "How to Use The CD" topic, we configured a path to the *C:\MasterVIZr\Maps* folder installed with the dataset from the Mastering VIZ Render CD ROM. (If you have not installed the VIZr datasets from the CD, please do so before configuring this path). Let's verify that this path is configured correctly now and add another for the RPC content provided with Mastering VIZ Render.

Verify that the *C:\MasterVIZr\Maps* folder is listed. If it is not, use the process in the next two steps to add it.

3. Click the Add button on the right to add a path to the list.

4. In the Choose New Bitmap Path dialog, browse to the *C:\MasterVIZr\Maps* folder, and then click the Use Path button (see Figure 9–2).

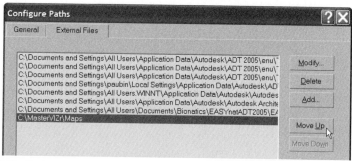

Figure 9–2 *Verify the path to the Maps folder*

5. Click OK to exit the Configure Paths dialog.

 TIP If you share your VIZ Render scenes with another person, always be sure to also send them the required external files as well. Adding a path as we have done here will only be useful if the folder named in the path actually contains the required bitmap files.

Add an RPC

Now that we are certain that the required path is configured, let's add our first RPC.

6. Maximize the Top viewport.

7. From the RPC menu, choose **RPC**.

NOTE *The free RPC Plug-in is required to continue. Please read the "Installing RPC" sidebar above for more information on how to download and install it.*

In the Modify panel you will see the RPC controller. This will change depending upon what type of RPC you have selected to insert, and also depending on which items you have installed. If you do not see any items listed, do the following:

 NOTE If you already have RPC installed, and the Environments group is showing, skip the next two steps.

8. Click the Configure Content button.

   This will open the RPC Content Configuration dialog.

9. On the Update tab, click the Update Content button and then close the dialog.

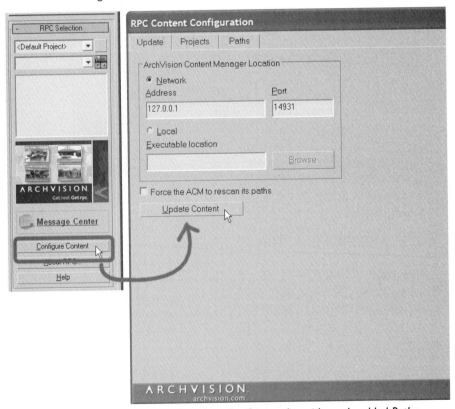

Figure 9–3 *Use Configure Content to update the Content list with newly added Paths*

You should now see an Environments list beneath the "Project Pulldown" list. Beneath that, in the "Content Category," should be listed an item named **Environ-**

ment01. If you already had RPC installed and have other RPC content, please choose Environments from the Content Category list and then select **Environment01**.

10. Click a point somewhere in the Top viewport and begin to drag. (Don't click the second point yet.)

The exact location is not critical; we will move it in an upcoming step.

As you drag the mouse, you will notice that the RPC is now rotating around the insertion point. The next point you click will be the location of the sun in the image that makes up the sky in the Environment RPC.

11. Move your mouse to about the 5 o'clock position and then click.

12. Right-click the mouse to end the creation mode.

Modify the RPC

13. With the RPC still selected, right-click and choose **Move** from the Edit/ Transform Quad menu.

14. In the Status Bar, type X: **20'-0"**, Y: **30'-0"**, Z: **-50'-0"** (see Figure 9–4).

Figure 9–4 *Center the RPC on the scene and lower it in the Z a bit*

This centers the RPC on the Retreat Pavilion.

 NOTE We are placing the RPC at -50'-0" in order to accentuate the fact that this box is on a hill.

Please note that the Environment RPC will only show in a rendered view. It does not show in the shaded viewport. Another important fact to note about RPCs is that they don't receive daylight very well. Their self-illumination setting must be set very high in order for them to not appear washed out when you render. In this particular Environment RPC, the included Sky and Horizon actually do quite well, but the Ground Plane needs some adjustment.

15. With the RPC still selected, locate the RPC Edit Tools rollout in the Modify Panel, and then click the Mass Edit button.

16. In the RPC Mass Edit dialog, click on the 0.0 under the light bulb icon.

17. In the right-hand pane, set the Self-Illumination value to **100**, click Apply, and then click OK to close the dialog (see Figure 9–5).

Figure 9–5 *Set the Self Illumination of the Environment RPC in the Mass Edit dialog*

18. Save the scene.

The Environment RPC utilized here has been provided free of charge by ArchVision for readers of *Mastering VIZ Render: A Resource for Autodesk Architectural Desktop Users*. Many additional RPCs are available for purchase. For more information, please visit *www.archvision.com*.

CAMERA ANIMATION

Since we have already explored many rendering and radiosity settings in previous chapters, in this sequence we will focus on the animation settings and use a provided Render Preset optimized for this first animation.

Test Rendering

Before we proceed to the animation, let's perform a quick test rendering to see how the image will look.

1. Set the Camera – Animation viewport active.

TIP Simply press the '**C**' key on the keyboard.

2. From the Rendering menu, choose **Render**.

3. At the bottom of the Render dialog, load the *Ch09_Draft Render_First Pass* Render Preset.

TIP If you wish, you can change the path to your Render Presets folder. Choose Configure Paths from the Customize menu, and then, on the General tab, modify the Render Presets path to point to the *C:\MasterVIZr\Render Presets* **folder.**

4. In the Select Preset Categories dialog, make sure everything is selected and then click Load.

5. Click the Render button at the bottom of the dialog (see Render Stats 09.01).

Render Stats 09.01

Render Time: 45 seconds
Radiosity Processing Time: 20 seconds
Music Selection: "Driven Under" Seether – Disclaimer (2002)
Color Plate: NA
JPG File: Rendering 09.01.jpg

You will notice an extra dialog named RPC Preparation before the rendering processes. This shows you the progress of the RPC calculation for the rendering (see Figure 9–6).

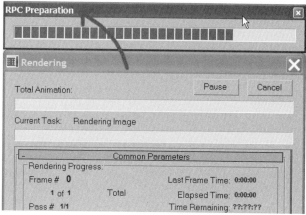

Figure 9–6 *An additional progress bar appears for the RPC Content*

The rendering that we just made is a 600 x 600 pixel version of what we will be animating shortly. However, in our first animation study, we will use a much smaller resolution.

Animate the Camera

In order to make a smooth turn around the building we will animate the rotation of the camera about the center of the building. If we were using VIZ or 3ds max, we could simply create a circular spline and attach the camera to it as a path. In VIZ Render, we are unable to do this, so we need to be a bit more creative.

6. Using the Selection Floater, select the Camera named Camera – Animate (close the Selection Floater when you're done).

7. On the Modify Panel, in the Parameters rollout, change the Type to **Free Camera**.

A target camera will not allow us to rotate about a point, since it is tied to its Target.

8. Switch to the Top viewport and Maximize it.

NOTE If you still have a maximized Camera viewport active from the previous sequence, simply press the '*T*' key on the keyboard.

9. With the Camera still selected, right-click and choose the **Affect Pivot only Mode Toggle** from the Edit/Transform Quad menu.

A large wireframe axis tripod will appear on the Camera. This is the Pivot Point of the Camera. The Pivot Point is the point upon which an object will rotate or transform. Moving the Pivot Point of the Camera will allow us to control the point around which it rotates.

10. Right-click again and choose **Move** from the Edit/Transform Quad menu.

11. Using the Y-axis constraint of the Move Gizmo, move the pivot point up to the center of the building (see Figure 9–7).

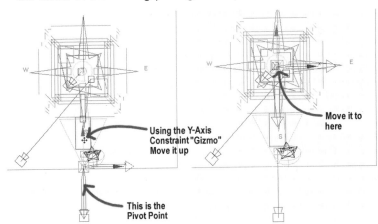

Figure 9–7 *Move the Pivot Point of the Camera up to the Center of the Retreat*

12. Right-click one last time and choose the **Affect Pivot only Mode Toggle** again to exit this mode.

This will hide the Pivot Point again, and the next time we use any of the transforms such as move or rotate, the Camera will actually transform itself—not just its Pivot Point—as we did here.

Time to Animate

Along the bottom of the screen is the Time Slider (see Figure 2–1 in Chapter 2). This tool is used to coordinate what happens and when it occurs. Animation in VIZ Render involves a fairly simple process—turn on the Animate (Auto) button, move the time slider to a point in time, then move an object. That object will then move from its original position to the new position between the start time and the time set on the slider when you render the animation. Much like old-style "cell" animation, a key frame is created at the point in time where you set the Time Slider. VIZ Render will interpolate the frames in between each key frame when you produce your animation.

13. Click the Animate button (labeled "Auto" at the bottom-right side of the screen—see Figure 9–8).

The Time Slider will turn red as an indication that you are in animate mode.

Figure 9–8 *Turn on Animation mode by clicking the "Auto" button*

14. Move the time slider to **50** (see Figure 9–9).

Figure 9–9 *Move the Time Slider to set a Key Frame*

The time slider is set to 100 frames by default. Animated movies run at a rate of 24 frames per second. At the end of this section we will have less than 4.2 seconds of an animation at that rate. For now we will work with this default, and later on in this chapter we will look at ways of changing this setting.

Be sure the Camera object is still selected.

15. Right-click in the Top viewport and choose **Rotate** from the Edit/Transform Quad menu.

16. On the Transforms toolbar, right-click the Angle Snap Toggle.

17. In the General area, change the Angle (deg) setting to **45**, and then close the Grid and Snap Settings dialog (see Figure 9–10).

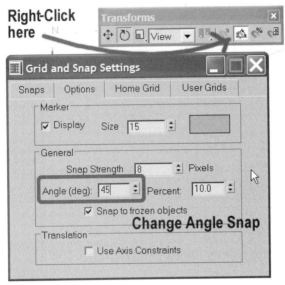

Figure 9–10 *Configure the Angle Snap to 45 degrees*

Move your mouse over the Z-axis gizmo (the blue dot pointing out from the screen). A Rotate cursor will appear.

18. Click and hold down the mouse with the rotate cursor over the Z-axis.

19. Rotate the camera **180°** clockwise (see Figure 9–11).

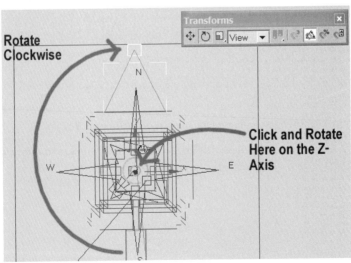

Figure 9–11 *Rotate the Camera Clockwise 180° around the Z Axis*

Notice it is rotating around the point where we put the Pivot Point in the steps above.

20. Move the time slider to **100**.

21. Rotate the Camera back to its original position (South), but continuing in the clockwise direction.

22. Turn off Animate (click the Auto button again).

    The Time Slider will no longer be highlighted in red.

Play the Animation in the Viewport

23. Make the Camera – Animate viewport active (press the '**C**' Key).

To the right of the Animate button there is a series of controls like the ones on a typical DVD player (see Figure 2–1 in Chapter 2). We use them here in VIZ Render in the same way that we would on our DVD player.

24. In this group of animation controls, click the Play button.

25. Click the same button again (which has now changed to a Pause button) to stop the animation.

Render the Animation

Playing the animation in the viewport is helpful in understanding if the key frames are working properly and to get a general idea of how the animation will flow. To really know what the result will be, we need to render the animation and create a digital file. In this exercise, we will actually render each frame to a separate file and later compile it into a video file. This will take a bit longer than rendering has taken until now. Remember, in all of the previous chapters, we have rendered a single still image—the equivalent of one frame in the animation! Imagine if we did an animation of the final rendering in Chapters 7 or 8. It could take several hours for all the frames to render. Fortunately, there are plenty of measures we can employ to speed up rendering time for animation.

The easiest and most significant way to reduce the time it takes to render is to reduce the size of the output file. Above we rendered a 600 x 600 pixel still image in approximately 45 seconds. However 20 seconds of that time was spent processing the Radiosity. Therefore, the rendering itself really took about 25 seconds. It turns out that Radiosity only needs to be calculated once for the scene and then that solution is used for all subsequent frames. If we rendered all 100 frames of this animation at the same size, with the same settings, it would take approximately 100 x 25 seconds, plus the additional 20 seconds for the Radiosity solution, or 2,520 seconds (which equals 42 minutes). This is not necessarily a tremendous amount of time, but for our first animation, we want to get a little closer to instant gratification.

At the pixel dimensions used above, we have a total of 360,000 pixels (600 x 600) in our image. Each pixel must be generated by the rendering engine and therefore, the more pixels, the longer it takes to render. It is simple math. What would happen if we dropped the size to 200 x 200 pixels? This is only 40,000 pixels; or one-ninth of the size of the original. Therefore, it stands to reason that our rending should take significantly less time—albeit maybe not exactly one-ninth the time. There are other settings that we can adjust as well to help reduce the time even further. Naturally your exact times will vary from ours depending on the specifics of your hardware, but the principals should remain the same.

26. From the Rendering menu, choose **Render**.

27. In the Output Size area, change both the Width and Height to **200**.

28. In the Render output rollout, click the Files button.

    Your browse window should already be in the *Chapter09* folder. If it is not, browse there now.

29. Double-click the *Animation* folder and then the *Frames – Camera* folder.

30. In the File Name field, type the name **Camera-**, and then click Save (see Figure 9–12).

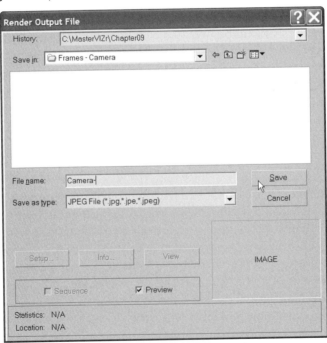

Figure 9–12 *Type a partial name for the root of each frame's file name*

31. In the JPEG Image Control dialog, simply accept the defaults by clicking OK.

A single file will be created for each frame. This file name input here will become the root file name of each frame generated. Appended to this will be a four digit suffix starting at 0000. As a result, the first rendered frame will be *Camera-0000*, and the last one in this particular series will be *Camera-0100* (since we have 100 frames).

There is one last setting—we need to tell VIZr which frames to render. In this case we will be rendering all the frames in the sequence. However, we have multiple options, one of which is to render every *nth* frame. This option is quite handy in a long animation, to get a sense of the entire animation without the expense of rendering each frame. For example, you might choose to render every third frame to create a quick test render.

32. Scroll back up to the top to the Common Parameters rollout in the Time Output area, and select the Active Time Segment radio button.

33. Leave the Render Scene dialog open. From the File menu, Save the scene.

34. In the Render Scene dialog, verify that the Camera – Animate viewport is active in the Viewport list at the bottom, and then click the Render button (see Render Stats 09.02).

 NOTE If you changed your view in the scene then a different view will be active in the Render dialog. Change your view in the scene and it will change in the Render dialog as well.

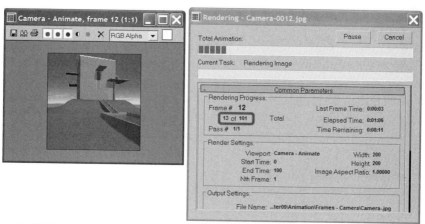

Figure 9–13 *Follow the progress of each frame*

As each frame renders, you can watch the progress bars in the Rendering window. You may notice that Radiosity processing occurs only at the beginning. As was noted above, once the Radiosity solution has been processed, it will be used for all frames. The only exception would be if the lighting were changing over the course of the animation (see Figure 9–13).

Render Stats 09.02

Render Time: 6 minutes 31 seconds (100 frames, 3.9 seconds per frame on average)
Radiosity Processing Time: 16 seconds
Music Selection: "Bohemian Rhapsody" Queen – Greatest Hits, Vols. 1-2 Disc 1 (1995)
Color Plate: NA
JPG File: Animation 09.02.avi

So your rendering is complete, and most likely you would like to see your completed animation. Since we have chosen to render each frame as a still image, we do not have a single animation file that we could load into something like Windows Media player. Later in this chapter, we will combine all of the separate stills into a single AVI file. However, we do not have to wait until then to preview the animation. In the next section, we will learn how to use the VIZr RAM Player to load all of the stills and play them as a continuous animation.

You may be wondering why we have not simply rendered the animation directly to an AVI file. While this approach is certainly possible, the approach advocated here gives us more flexibility. By rendering each frame, we have access to each file before we make the final rendering. Also, should a problem occur during the rendering process, such as a power outage, we would not lose any of the frames that had already been rendered. In the same vein, we could stop the rendering at any time and choose to resume it later, picking up at exactly the frame where we left off.

RAM PLAYER

RAM Player loads rendered frames into your computer's RAM (Random Access Memory). It has two channels labeled A and B so that you can load two renderings simultaneously and then compare them side by side. Later on we will create another animation and then load both it and the one we just completed into RAM Player at the same time and compare them. For now, however, let's open RAM Player and have a look at the animation that we just completed.

NOTE Performance with RAM Player naturally depends heavily on how much physical RAM you actually have on your system.

1. From the Rendering menu, choose **RAM Player**.

2. On the left side, click the Open Channel A icon (see Figure 9–14).

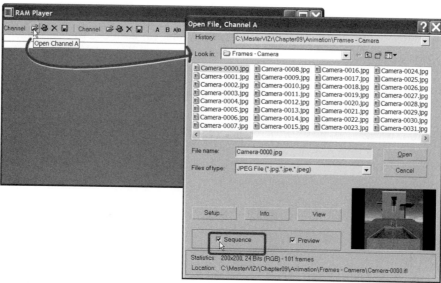

Figure 9–14 *Open Channel A in RAM Player*

3. Browse to C:\MasterVIZr\Chapter09\Animation\Frames – Camera.

In this folder, you will see all of the JPG files that were rendered above.

4. Select *Camera-0000.jpg*.

5. At the bottom of the Open File dialog, place a checkmark in the Sequence checkbox (see Figure 9–14).

By choosing Sequence, we will be prompted to indicate which frames we want to load. If you do not choose Sequence, you will only be loading a single frame.

6. Click the Open button, and then in the Image File List Control dialog, click OK to accept the defaults.

In the Image File List Control dialog are four options:

▶ **Start Frame**—Choose the first frame of the sequence that you wish to view.

▶ **End Frame**—Choose the last frame of the sequence that you wish to view.

▶ **Every n**th—Use this if you wish to skip frames during playback. This may be handy if you have rendered a lot of frames and you would like to gain a sense of the entire animation without waiting to load every frame. For instance, to see only every fourth frame, input 4 for this value.

▶ **Multiplier**—This setting will load each frame in the sequence the indicated number of times. This is quite useful for getting a sense of the entire length of the animation. For example, you may set the animation to render every 5th frame (using Every n th) to save on rendering time. In this dialog you can set the multiplier to 5, and you'll have the full number of frames that your final rendering will contain without the cost of waiting for each one to render. The animation will be choppy, but for test animations this is not a major concern.

7. In the RAM Player Configuration dialog that appears, click OK to accept the defaults.

Typically you will want to preview the files in RAM Player as they were rendered; therefore, even though it is possible to change the resolution here, it is not recommended. If you have enough physical memory on your system, you can increase the Memory allotment.

8. In the RAM Player, click the Playback Forward icon (see Figure 9–15).

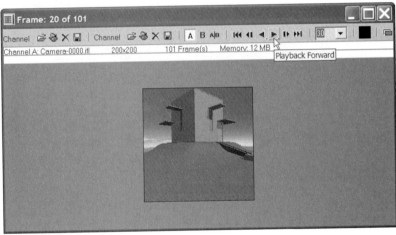

Figure 9–15 *Play the Animation in RAM Player*

If you have room on your monitor, leave RAM Player open onscreen for the next sequence. You can also Minimize it and then recall it later from the Windows Task Bar.

9. Save the scene.

RENDER A LIVE SECTION

It might be fun to generate another animation of the model as it appears in section. To do this, we'll go back to ADT and enable a Live Section. A Live Section shows a section cut through the model and removes the display of the sectioned portion. Optionally you can choose to render the sectioned portion in a different material, but for this exercise we will keep things straightforward.

Create a Live Section in ADT

1. Launch Architectural Desktop and open the *Retreat.dwg* file from the *C:\MasterVIZr\Chapter09* folder.

2. On the Design palette (in the Design Tool Palette Group), click on Vertical Section tool.

 NOTE Tool Palette Groups contain selections of Tool Palettes. In the out-of-the-box installation of ADT, there are three Tool Palette Groups: Design, Documentation, and Detailing. Right-click the Tool Palettes title bar to load a group.

In the next few steps, use your NODE Object Snap to snap to the points provided in the drawing.

3. At the "Specify start point" prompt, snap to the Node at the bottom-center of the model.

4. At the "Specify next point" prompt, snap to the Node at the top-center of the model (see Figure 9–14).

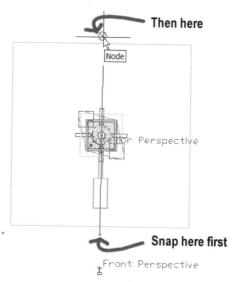

Figure 9–16 *Create a Section Line snapped to the provided Node Points*

5. Press ENTER to complete the section line.

6. At the "Enter length" prompt, type **65'-0"**, and then press ENTER.

This process will create a section line object that surrounds the left side of the model. This is the region that will be included within the Live Section. The rest of the model that lies outside of this boundary will be cropped away.

7. Select the Section Line, right-click, and choose **Enable Live Section**.

From the Top view we will see no change. Live Sections apply only to three-dimensional Display Configurations. Top view uses a two-dimensional Plan Display Set. To see the effect of the Live Section, we need to switch to a 3D view.

8. From the View menu, choose **Preset Views > SE Isometric** (see Figure 9–17).

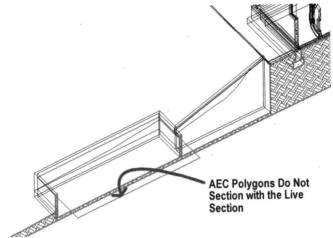

AEC Polygons Do Not Section with the Live Section

Figure 9–17 *The Live Section does not apply to AEC Polygons*

If you zoom-in you will notice that some objects were not cut by the Live Section. These objects are AEC Polygons. AEC Polygons are style-based but Material Definitions cannot be applied to them. This limitation precludes their effective use within Live Sections. For the purposes of our explorations here, the simplest way to deal with these objects is to simply not include them in VIZr.

9. On the Properties palette, click the Quick Select icon.

Quick Select makes a selection set based upon a criterion that you specify in its dialog. Since we want to select all of the AEC Polygons in the drawing, we will use Quick Select to select all of the objects within this drawing that are AEC Polygons.

At the top of the dialog, be sure that Apply to reads **Entire Drawing**.

10. From the Object type list, choose **AEC Polygon**.

11. Skip over the Properties field and choose **Select All** from the Operator list.

12. Be sure that Include in new selection set is chosen, and then click OK (see Figure 9–18).

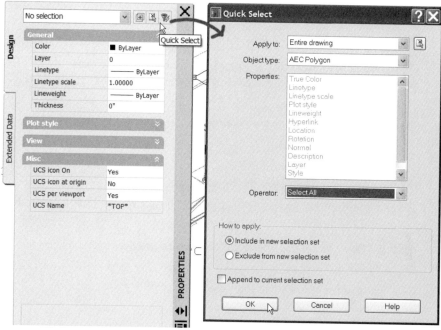

Figure 9–18 *Use Quick Select to select just the AEC Polygons*

13. Select also the Section Line object.

    You should now have all the AEC Polygons and the Section Line selected.

14. Right-click and choose Isolate Objects > Hide Selected Objects.

This tool hides objects based upon selection only—it is object-type- and layer-independent.

15. Save the ADT file, and then switch over to VIZ Render.

Save a New VIZr Scene

So that we preserve the work that we have done so far, let's save a copy of the current scene. Remember, making a copy allows us to explore different options while leaving both (all) copies linked back to the original ADT file.

 NOTE New in VIZ Render is the Scene States functionality. While not utilized in this tutorial, you could use this method to save the current state of the scene and then create a variation instead of the Saveas method used here. Search the online Help for more information on Scene States.

16. From the File menu, choose **Save Copy As**.

17. For the File name, type **Retreat – Live Section**.

18. From the File menu, choose **Retreat – Live Section**.

    If prompted, Save the current file.

19. On the file Link toolbar, click (turn on) both the "Use scene material definitions" and "Use scene material assignments (on Reload)" toggle icons.

 TIP If you need a refresher on the reload options, refer to Chapter 2.

20. On the File Link toolbar, click the Reload icon.

When the Reload is complete, half of your model will disappear. This is expected since we enabled the live section.

21. Click the Play icon on the Animation controls to preview the animation in the viewport.

    Click Pause when you are finished viewing.

Render the Live Section View

Now, let's render the animated camera for this version of the model. Everything is already set up for us to render the same Camera since we saved this file from the previous one.

22. From the Rendering menu, choose **Render**.

23. In the Render Scene dialog, click the Files button in the Render Output rollout.

24. Browse to the C:\MasterVIZr\Chapter09\Animation\Frames - Live Section folder.

25. For the File name type **Live Section-**.

26. In the JPEG Image Control dialog, simply accept the defaults by clicking OK.

27. In the Render Scene dialog, click the Render button (see Render Stats 09.03).

 ### Render Stats 09.03

Render Time: 6 minutes 15 seconds (100 frames, 3.9 seconds per frame on average)
Radiosity Processing Time: 6 seconds
Music Selection: "Bad" U2 – The Unforgettable Fire (1984)
Color Plate: C–14 (bottom)
JPG File: Animation 09.03.avi

COMPARING TWO ANIMATIONS IN RAM PLAYER

We will now return to RAM Player and look at the second animation. We will also look at both the original animation and the new Live Section one we just completed side-by-side for comparison.

> RAM Player should still be open onscreen with the Camera animation already loaded in Channel A. If you closed RAM Player, please repeat the steps in the "RAM Player" heading above to open it again and reload the animation in channel A.

Load the Live Section Animation in Channel B

1. In RAM Player, click the Open Channel B icon (see Figure 9–19).

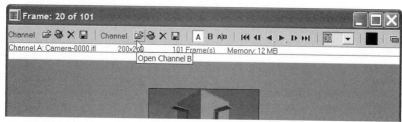

Figure 9–19 *Click the Open Channel B icon to load the second animation into RAM Player*

2. Browse to the *C:\MasterVIZr\Chapter09\Animation\Frames - Live Section* folder.

3. Select the *Live Section-0000.jpg* file, place a checkmark in the Sequence checkbox, and then click Open.

4. As we did above, click OK in the "Image File List Control" dialog, and then click OK again in the "RAM Player Configuration" to accept the defaults.

Once RAM Player has finished loading the frames, you will see two small arrows at the top and bottom of the image. Even though it looks like we are seeing the first frame of channel B, we are actually seeing half of channel A on the left and half of channel B on the right (see Figure 9–20).

 NOTE Notice that the two Channels list a file name with an "IFL" extension. An IFL (Image File List) file is an ASCII file that constructs an animation by listing single-frame bitmap files to be used for each rendered frame. We generated one when we chose Sequence and decided what the beginning and ending frames were. Refer to the VIZr help for more information.

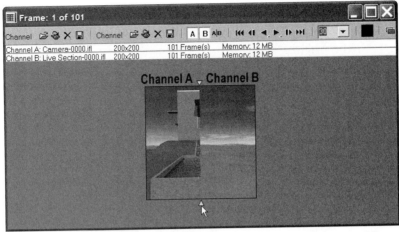

Figure 9–20 *RAM Player showing channels A and B split vertically down the middle*

Before we begin to explore this exciting functionality, let's turn off Channel A and play the Live Section animation.

> 5. On the RAM Player toolbar, click the 'A' icon (currently depressed) to turn it off (pop it up).

The small arrows on the top and bottom disappear and now only the 'B' icon is depressed, indicating that we are now seeing only Channel B.

> 6. Click the Playback Forward icon to preview the Live Section animation.

> 7. Click the Stop icon (the same icon changes to Stop while playing) when finished.

RAM Player Playback Controls

In addition to simple forward playback, we can preview our animation(s) a few other ways as well.

> 8. Click the Playback Reverse icon.

> Stop the Playback when you're satisfied.

As you can see, the animation will play in the opposite direction. In both directions, you can step forward or backward one frame at a time, or jump to the first or last frame (see Figure 9–21).

> 9. Try each of these icons now.

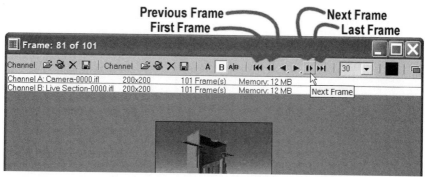

Figure 9–21 *Move forward and backward through each frame*

Perhaps you would like to preview the animation at a faster or slower speed. Next to the playback control icons is a dropdown list. This controls the playback speed represented in terms of frames per second; or its "frame rate."

10. From the Frame Rate list, choose **15**, and then click the Playback Forward icon.

11. Try other speeds while the animation is playing.

    Stop the Playback when you're satisfied.

Side-by-Side Comparison

Now that we have previewed both animations by themselves in the RAM Player, let's explore the ability to do side-by-side comparisons of two renderings in the same RAM Player window. This is handy when you are making changes to materials, lighting, or Radiosity. You can load two still images or animations into channels A and B, and then compare them to one another interactively.

12. Choose **15** from the Frame Rate list.

13. Click the 'A' icon to turn on Channel A again.

    The small arrows will reappear at the top and bottom.

14. Click the Playback Forward icon.

    Try Reverse Playback, and Next and Previous Frame as well if you wish.

15. Stop the animation at an interesting frame. (Use any of the playback controls to assist you.)

16. Click and drag one of the small arrows at the top or bottom of the image.

    Slide (scrub) it left and right (see Figure 9–22).

More of B   **More of A**

Figure 9–22 *Use the slider to "scrub" between Channel A and B*

17. Click the A|B toggle icon at the top of the RAM Player to switch to a horizontal split.

18. Scrub the divider again.

Let's now play both animations and scrub between them as it plays.

> **NOTE** As its name implies, RAM Player works best with lots of physical memory. If you are working on an older system, or do not meet the minimum system specifications for ADT and VIZr, you may experience problems with this sequence.

19. Click either Playback icon and then use the scrub slider to scrub between both animations as they play.

    Feel free to try any of the other tools: Forward, Reverse, Frame Rate, Horizontal, and Vertical toggle as you playback the animations.

SAVING AND MODIFYING ANIMATION

Once you are satisfied with your explorations in RAM Player, you may want to save a single animation file such as an AVI so that you can play it outside of VIZr. Next to each channel on the toolbar is a Save icon. You will not be able to save an animation as you see it in the split channels; you must save either Channel A or Channel B.

Save the Animation

20. Click the Save icon for Channel A.

21. Browse to the *C:\MasterVIZr\Chapter09\Animation* folder.

22. In the File Name field, type **Camera**.

23. From the File Type list, choose **AVI File**, and then click Save (see Figure 9–23).

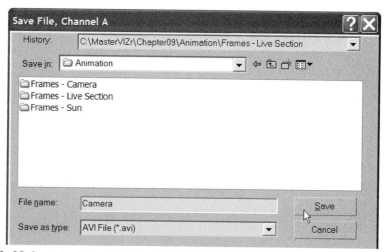

Figure 9–23 *Save the Animation in Channel A to an AVI file*

> 24. In the "AVI File Compression Setup" dialog that appears, accept the defaults and then click OK.

An option here that works for some professional-level animators is to save the individual high-resolution JPEGs to an uncompressed AVI. From there, this AVI is opened in programs such as Adobe Premier or After Effects for post-processing and conversion into a compressed final MPEG file. If you are not familiar with such procedures and do not have access to such tools (don't worry, we don't either), the defaults will yield fine results.

> 25. Repeat the steps for Channel B and name the AVI file **Live Section**.
>
> 26. Close RAM Player.

When you close the RAM Player, VIZr will ask if you want to clear the RAM. Since we saved these files into RAM, and adding them took some time, VIZr wants to be sure that you want to do this (see Figure 9–24).

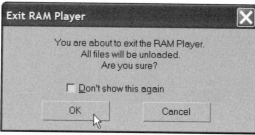

Figure 9–24 *Exit RAM Player Warning*

> 27. Click OK in the Exit RAM Player warning dialog.

28. In Windows Explorer, browse to the *C:\MasterVIZr\Chapter09\Animation* folder and double-click either *Camera.avi* or *Live Section.avi* to open them in Windows Media Player (or any other media player that can play back AVIs) and see the results.

Since we were concerned about the time it takes to render, we rendered these animations at a very small size. This was discussed in detail above. Now that you understand the process, feel free to go back and repeat this process but render the animation at a higher resolution, such as 600 x 600 pixels. It will take much longer to create the animation, but the larger size will be much more gratifying.

MODIFYING THE RENDERED FRAMES

There are two ways of changing the time length of an animation. You can "tack on" frames to the beginning or end of the initial set, or you can "stretch out" what you have now. We will not render these here, but this information should prove useful in your animation endeavors.

TIP If you decide to try these changes now, save the scene first so that you can reload it later without saving the new animation times.

If you have objects in your scene that have been animated to move from a given point to another in the animation but you want there to be more time before or after the movement, tacking on frames is what you want to do. On the other hand, if you have an object that moves continuously throughout your animation but the animation is too short, the stretch method is the right option.

Tacking on Frames to an Existing Animation

This procedure adds new frames to the end of your animation, without affecting your existing work.

1. In the bottom-right corner of the VIZr screen, click the small Time Configuration icon (see Figure 9–25).

Figure 9–25 *Click the Time Configuration icon to Edit Animation Time*

2. In the Time Configuration dialog, in the Animation area, enter the number of the last frame of the animation in the End Time field or the first frame in the Start Time field.

For example, if your existing animation is 100 frames long and you want to add 50 frames, enter **150**. Your existing animation will be unchanged, but there will now be 50 additional frames at the end. The Time Slider at the bottom edge of the screen will update immediately. If you wish to add frames to the beginning of the animation, you can input a negative number in the Start Time field.

3. Click OK to dismiss the Time Configuration dialog.

Stretch an Existing Animation Over a Longer Time Period

If you want to play the existing animation over a longer period of time, use this procedure.

1. In the bottom-right corner of the VIZr screen, click the small Time Configuration icon (see Figure 9–25).

2. In the Time Configuration dialog, in the Animation area, click the Re-scale Time button.

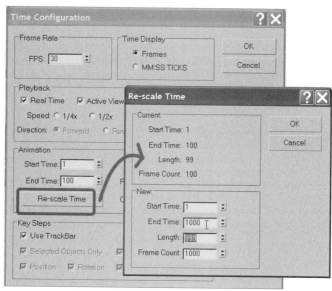

Figure 9–26 *Re-scale Time to "stretch" the existing animation over a longer period of time*

3. Change the values in any of the fields in the New area.

For example, if you wanted the animation to run for 1,000 frames rather than 100, you could input the Start Time as 1 and the End Time as 1,000, or simply input 1,000 for the Frame Count.

4. Click OK twice to make the change.

If you play the animation, it will run the same as before, only now it will basically be slower. If you rendered at this new length, it would, however, create 1,000 frames—which would take considerably more time!

PANORAMA RENDERING

We now have some nice rendered animations of the exterior to show our client. Let's move to the inside and capture a rendering that can be viewed in 360 degrees horizontally and vertically interactively. This is called a "Panorama" rendering.

Save a New VIZr Scene

1. From the File menu, choose **Retreat – Animation Study** from the recent file list.

   This will load the file that we built our first animation from before we cut the Live Section.

2. From the File menu, choose **Save Copy As**.

3. For the File Name, type **Retreat – Panorama**.

4. From the File menu, choose **Retreat – Panorama** from the recent file list to load the new file.

5. Minimize the viewports.

Add and Position a New Camera

6. Add a Target Camera anywhere in the scene, and name it **PanoramaCam**.

   Right-click to avoid adding additional Cameras.

7. With **PanoramaCam** still selected, right-click and choose **Move** from the Edit/Transform Quad menu, and then right-click again and click the small Transform Type-In icon next to Move (see Figure 9–27).

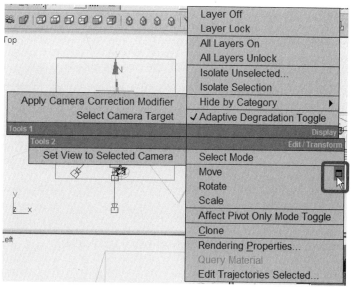

Figure 9–27 *Open the Transform Type-In*

8. In the Absolute:World area of the Move Transform Type-In, change the coordinates of the Camera to X:**20'-0"**, Y:**30'-0"**, Z:**8'-0"**.

9. Leave the Move Transform Type-In open, and then, on the Modify Panel, set the Lens to **15mm** (from the Stock Lenses in the Parameters rollout).

10. Right-click and choose **Select Camera Target** from the Tools 1 Quad menu.

11. In the Absolute:World area of the Move Transform Type-In, change the coordinates of the Target to X:**20'-0"**, Y:**50'-0"**, Z:**8'-0"**.

12. Close the Move Transform Type-In, and then Save the scene.

For more information on adding Cameras, refer to Chapter 5.

Adjust the Environment for an Interior Scene

Now that we've moved inside, we need to turn off the Exterior Daylight setting and brighten up the scene.

13. From the Rendering menu, choose **Environment**.

14. In the Logarithmic Exposure Control Parameters rollout, remove the checkbox from the Exterior Daylight checkbox.

15. In the same rollout, change the Brightness to **45**.

16. Click the Render Preview button (up and to the right) to make sure these values are satisfactory.

17. Close the Render Scene dialog.

18. On the Utilities Panel, choose Panorama Exporter.

19. In the Panorama Exporter rollout, click the Render button.

This will open a Render dialog similar to the Render Scene dialog. However, this dialog has only one tab. It will grab the other required settings from the standard Render Scene dialog.

20. From the Viewport dropdown at the bottom, choose **PanoramaCam**, and then click the Render button (see Figure 9–28 and Render Stats 09.04).

Render Stats 09.04

Render Time: 45 seconds (about 7.5 seconds per frame)
Radiosity Processing Time: 20 seconds
Music Selection: "Forever And Always" Shania Twain – Up! (2002)
Color Plate: NA
JPG File: Rendering 09.04.jpg

Figure 9–28 *Render the PanoramaCam in the Render Setup Dialog*

View the Panorama

When the Panorama Exporter Viewer dialog opens you are able to use your mouse to navigate the rendering. Simply click with the mouse and move in the direction you want to look: N, S, E, and W. Small mouse movements move more slowly, while larger mouse movements move more quickly.

If you would like to render a higher-resolution version, repeat the process and change either the Width or Height of the Output Size. If you input the Width, the Height will automatically adjust to half this size. Likewise, if you input a Height, the Width will automatically adjust to twice the Height. This proportion is required to create a Panorama rendering. This is because it actually creates four renderings wide (for the front, back, and sides) by two high (for the top and bottom).

If you would like the same level of quality that you got with your still renderings, load one of the presets provided for Chapter 9 and re-render the panorama. Two Render Presets are included in the *C:\MasterVIZr\Chapter09\Presets* folder: *Ch09_Panorama Render_Mid Range.rps* and *Ch09_Panorama Render_High Res.rps*. Naturally, the "High Res" one will yield the best results, but will also take the most time to render. Sample panoramic renderings from each Preset are included on the

CD ROM. Each was rendered at 1,600 x 800 pixel resolution (see Render Stats 09.05 and 09.06).

 ### Render 09.05

Render Time: approximately 6 minutes (about 1 minute per frame)
Radiosity Processing Time: 4 minutes 7 seconds
Music Selection: "Rapture" Blondie – The Best Of Blondie (1989)
Color Plate: NA
JPG File: Rendering 09.05.jpg

 ### Render 09.06

Render Time: approximately 30 minutes (about 5 minutes per frame)
Radiosity Processing Time: 4 minutes 7 seconds
Music Selection: "Cling" Days Of The New – Days Of The New (1997)
Color Plate: C–14 (middle)
JPG File: Rendering 09.06.jpg

If you wish to save your panoramic renderings, you can click the Files button in the Interactive Panorama Exporter Output rollout and choose one of many common file formats, such as QuickTime VR. Your recipients will need an appropriate viewer (usually free to download) to view the Panorama files.

SUN SHADOW STUDY

The final exploration covered in this chapter is to present the model in a more schematic look. We want to explore the shadow play of the model on the landscape and the various awnings and sunshades against the façades of the retreat.

White Strathmore Study Model

The best way to study the shadows will be to simplify the materials. We can easily swap the materials used in this model for a simple Strathmore-type material. A simple white Strathmore material will not detract from the shadow play like other materials may.

1. From the File menu, choose **Retreat – Animation Study** from the recent file list.

2. When prompted, Save the *Retreat – Panorama* file.

3. From the File menu, choose **Save Copy As**, and then type **Retreat – Strathmore** for the name.

4. From the File menu, choose **Retreat – Strathmore** to load this file.

5. On the Scene – Unused palette, right-click on the Create New tool, and choose **Paper** from the list of templates.

6. Right-click the new material and choose **Properties**. Change the name to **MVIZr – Strathmore**.

7. Change the diffuse color to pure white (R=**255**, G=**255**, B=**255**).

   Leave the remaining settings as is.

8. Open the Selection Floater.

9. In the text filter field at the top, type ***glass**.

Notice that this will highlight items in the list of objects as you type. The "*" is the standard Windows wildcard for "everything." When you first type *, all objects will highlight. As you type more of the letters, the highlighted items will shift to match the pattern. The pattern that we are using here indicates that the name can include any prefix, but must end with the letters "glass." This filter will isolate all of the Glass components.

10. Click the Invert button, and then click the Select button. Close the Selection Floater.

This will select all objects in the scene except those that are Glass.

11. On the Scene - Unused palette, right-click on the MVIZr – Strathmore material and choose **Apply To Selected**.

Notice that all of the materials that were originally applied to the selected objects have moved to the Scene - Unused palette in response to this change. This is because they are no longer assigned to the objects in the scene.

12. Set the Camera:SW Corner Perspective camera active.

   Maximize the viewport if it is not already maximized.

13. Orbit the camera so that you are looking down on the model (see Figure 9–29).

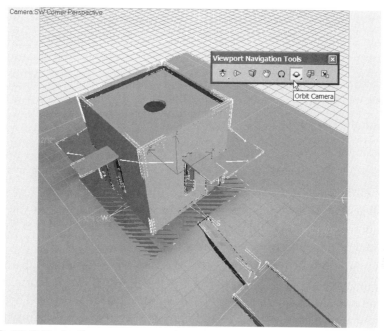

Figure 9–29 *Orbit the Camera to an aerial viewpoint*

Animate the Sun

We could perform a still rendering at this point to study the shadow play at a single point in time. However, it will be more valuable to watch the shadows move across the landscape. Therefore, let's work up another animation.

14. Select the *Daylight01* object and rename it **Animated 8-8**.

15. On the Modify Panel, with *Animated 8-8* selected, set the Hours to **8** in the Control Parameters rollout.

16. Click the Animate button.

   The Animate (Auto) button and the Time Slider will turn red.

17. Move the Time Slider to frame **100**.

18. On the Modify Panel, in the Control Parameters rollout, set the Hours to **20**.

19. Click the Animate button again to turn it off.

The effect of these settings will be that the time from 8am to 8pm will be divided by 100 (because we have 100 frames), and at each frame the sun will progress that many minutes. Here's the math:

8am to 8pm = 12 hours x 60 min/hr = 720 min/ 100 frames = 7.2 min per frame.

Therefore, 7.2 minutes of time will pass with each frame. A hundred frames at 24 frames per second gives us 12 hours of daylight in less than 4.2 seconds! That may be a bit too fast—let's take a look.

20. Minimize the viewport if you have one active so that you can see all four viewports.

21. Click the Play button to play the animation in the viewports.

22. Right-click in each of the viewports while the animation plays.

This will allow you to see the animation of the daylight system from the different angles.

Doing this allows you to see from where the sun is rising and setting. If you want the sun to fully rise and fully set, you will need to adjust the time of day at the beginning and end of the animation.

You will also notice that there is a camera flying around your building. That is the camera we animated earlier and saved to the file. It will not affect the current animation in any way, but later you might want to generate an animation for that camera again. If you did, you would be traveling 360° around the building over a span of 12 hours as the sun passed through the sky.

Render the Sun Study Animation

This animation will take longer than the other one because Radiosity will have to be calculated for each frame. This is because as the sun moves, it changes the lighting of each frame.

23. Set the *Camera:SW Corner Perspective* camera active.

24. From the Rendering menu, choose **Render**.

25. Load the *Ch09_Animation_Study.rps* Render Preset, or, if you prefer, configure the following settings:

   ▶ Verify that Active Time Segment is chosen in the Time Output area.

   ▶ In the Render Output rollout, click the Files button.

   ▶ Browse to the: *C:\MasterVIZr\Chapter09\Animation\Frames – Sun* folder.

   ▶ In the File name filed, type **Sun-**.

   ▶ From the Files of type list, choose JPEG File and then click Save.

26. Save the scene before you render.

NOTE Make it a habit to remember to save before rendering.

27. In the Render Scene dialog, click the Render button (see Render Stats 09.07).

Render Stats 09.07

Render Time: 25 minutes 15 seconds (100 frames, 16 seconds per frame on average)
Radiosity Processing Time: 12 seconds per frame on average
Music Selection: Eric Clapton - The Cream of Clapton, (1995) Entire CD
Color Plate: C–13
JPG File: Animation 09.07.avi

ADDITIONAL EXERCISES

When the rendering is complete, you will have 100 JPG files in the *Animation\Frames – Sun* folder. Using the same techniques that we used above in the "RAM Player" and "Comparing Two Animations in RAM Player" topics, you can view and save the animation. If you wish to let the rendering process overnight, you can increase the size of the frames and render it again.

The file that we have currently loaded would also make some interesting still renderings. Play the animation and use it to help you find your favorite time of day. Render high resolution stills of this model with its Strathmore material from orthographic Top, Front, and Right Views. This will give a more artistic look to these renderings. You can use your page layout program of choice to compose these renderings in a pleasing presentation.

You may also want to consider rendering the Live Section from the East. This is the view looking directly into the Live Section cut. You can do this one with or without materials.

On some of these renderings, you can speed them up by adjusting the Render Properties of the large Mass Element ground plane. Since there are already a lot of faces in the Mass Element, we can save time by excluding this from the normal Radiosity meshing process. To do this, select the Site Mass Element, right-click, and choose Rendering Properties. Uncheck Use Global Meshing, and also uncheck the Subdivide checkbox.

CONCLUSION AND SUMMARY

As you can see from this chapter's exploits, there are many ways that we can present our VIZ Render scenes for consumption by our peers and clients. With options for animation, live sections, white study models, and panoramas, you are sure to find the option that is right for the message that you wish to convey. In this chapter, we learned the following:

You can animate the Camera to move over time.

Live Sections can be created in ADT and then viewed, rendered, and animated in VIZr.

The RAM Player is a very useful and powerful tool for composing animations and comparing two renderings side by side.

RAM Player can be used to compile an animation file from a series of still renderings.

You can add frames, change the length of an animation, and also re-scale the frames to a new total length.

Sometimes you can convey more with less—swap all scene materials for simple materials like white Strathmore and clear glass.

Animating the Daylight System is an excellent design and presentation tool.

Study shadow play by animating the Daylight System.

Create a Panoramic rendering of an interior space to give the feel of being there.

Appendices

This section contains four appendices to the materials in this text. You will find additional exercises, online resources, render tables and render time stats included in these appendices.

Section III is organized as follows:

Appendix A Additional Exercises

Appendix B Online Resources

Appendix C Render Settings

Appendix D Render Times

Additional Exercises

INTRODUCTION

This Appendix includes several practice exercises to give you more experience with the topics covered in many of the chapters. It is intended that you will visit this appendix at the completion of each chapter. Once you have finished the lessons in a particular chapter, perform the exercises herein for that chapter. You are encouraged to experiment in each of these exercises. The notes given here are merely guidelines for your further explorations. Feel free to perform other tasks not listed and experiment with other tools. You can use SaveAs or Save Copy As to explore alternatives. Each additional exercise assumes that you will start where you left off at the end of the designated chapter unless otherwise noted. Some of these renderings will produce long render times. It might be a good time to get that copy of the "The Lord of the Rings - The Return of the King (Platinum Series Special Extended Edition) (2003)" on DVD. That will certainly help you pass the time.

CHAPTER 1

RETREAT

Here are some additional exercises to practice topics covered in Chapter 1 – Getting Started.

Higher Quality Render

Render with higher quality presets.

1. From the Rendering menu, choose **Render**. At the bottom of the dialog, choose **Load Preset**, browse to C:\MasterVIZr\Render Presets, and choose Ch01_Final Render_High Res.rps. Select all items and click the Load button.

2. On the Common tab, click the Files button. Change the "Save as type" to either JPG or PNG. Next, browse to a location such as C:\MasterVIZr\Chapter01 to save the file, input a name, and then click Save.

3. Render the scene.

This rendering may take several hours to render, so plan accordingly.

Render an Alternate Camera Angle

Create a new camera angle.

4. In the Camera01 viewport, press the '**P**' key.

5. Use the Viewport Navigation tools to view the interior scene from a new angle.

If you have not read Chapter 2 yet, you can find tips on using the Viewport Navigation Tools there.

6. When you are happy with the new view, hold down the CTRL key and then press **C** to create a new camera from the current perspective view.

7. Render the scene using any of the presets from this chapter.

CHAPTER 2

Feel free to experiment further with any of the tools covered in this chapter. The goal is to get comfortable with the VIZr user interface. In addition, be sure to look at the VIZ Render User Reference on the Help menu. You can use the Index or Search functions to explore any of the topics covered in Chapter 2. The help for VIZ Render is quite good, and we are certain that you will find it a valuable resource.

CHAPTER 3

Since Chapters 3 and 4 are two parts of the same exercise, there are no suggested additional exercises for Chapter 3. Please proceed to Chapter 4 directly.

CHAPTER 4

The model at the end of Chapter 4 is nearly complete. However, later when we render it, it may be noticeable that there is no back to the townhouse model.

Be sure that the *Townhouse04* Project is active in ADT.

1. Open the *Basement* Construct, select one of the side Walls, and choose **Add Selected**. Add a Wall along the back.

2. Perform any required cleanup.

3. Repeat this process for the *First Floor* and *Second Floor* Constructs.

Color Plates

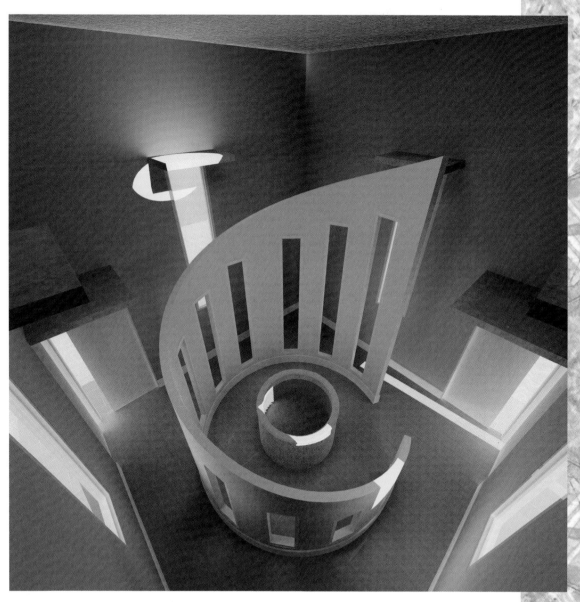

Figure C–1 *Chapter 1—Final Rendering using the Ch01_Final Render_High Res Render Preset*

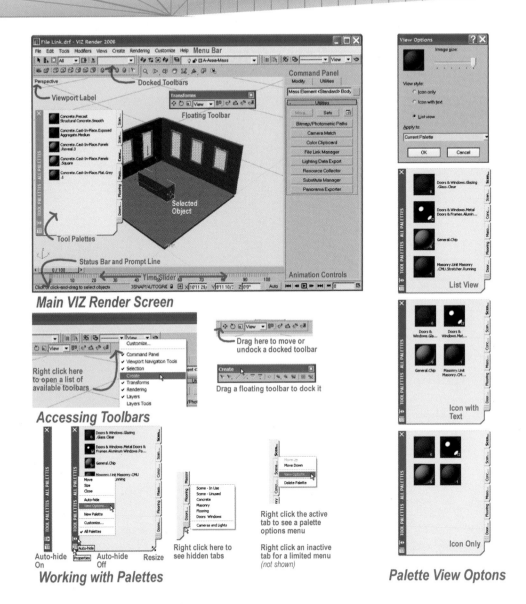

Figure C–2 *Chapter 2—The VIZ Render User Interface*

Figure C–3a *Chapter 2—Rendering the small room with Radiosity*

Figure C–3b *Chapter 2—Insert i-drop Content from manufacturer's content*

Herman Miller Web Site and i-drop Content used with permission from Herman Miller, Inc.
Scott Onstott Web Site and i-drop Content used with permission from Scott Onstott

Figure C–4a *Chapters 3 & 4—Details of the Townhouse Façade Model shaded in ADT*

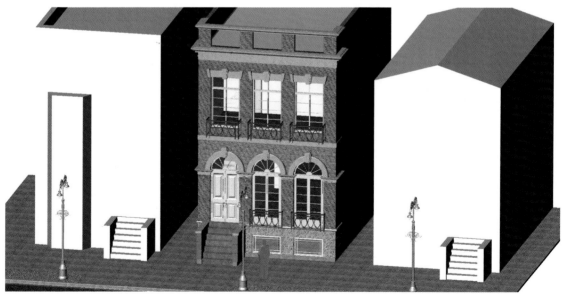

Figure C–4b *Chapters 3 & 4—The completed Townhouse Façade Model shaded in ADT*

Figure C–5 *Chapter 5—Final Rendering using the Ch05_Final Render_High Res Render Preset*

Figure C–6 *Chapter 6—Final Rendering using the Ch06_Final Render_High Res Render Preset*

Figure C–7 *Chapter 6—Alternate Exercise Final Rendering using the Ch06_Final Render_High Res Render Preset*

Figure C–8 *Chapter 7—Final Rendering using the Ch07_Final Render_High Res Render Preset*

Figure C–9a *Chapter 7—Townouse Project using the Ch07_Final Render_High Res Render Preset*

Figure C–9b *Chapter 7—Detail Renderings of the Townhouse using the Ch07_Final Render_High Res Render Preset*

Figure C–10 *Chapter 8—Final Rendering using the Ch08_Final Render_High Res Render Preset*

Figure C–11 *Chapter 8—Alternate Exercise Final Rendering using the Ch08_Final Render_High Res Render Preset*

Figure C–12 *Chapter 9—Final Rendering using the Ch09_Final Render_High Res Render Preset*

Figure C–13 *Chapter 9—Final Rendering using the Ch09_Final Render_High Res Render Preset*

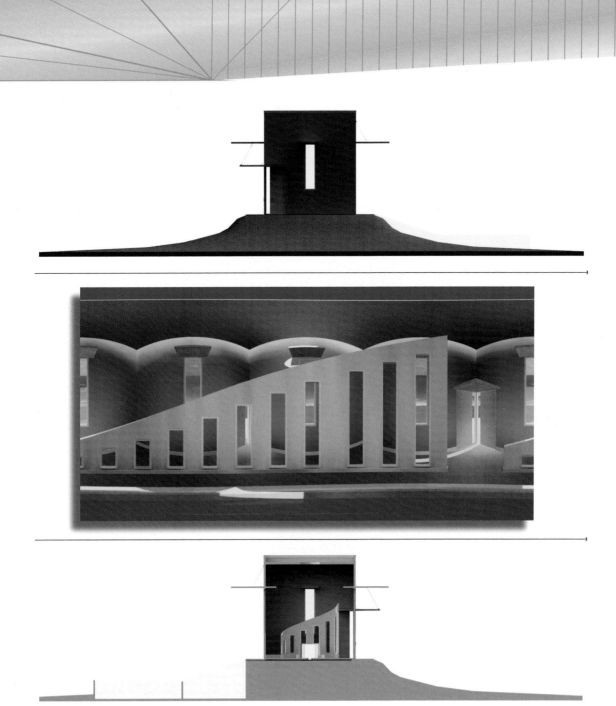

Figure C–14 *Chapter 9—Orthographic, Panoramic and Live Section Renderings of the Retreat*

Figure C–15a *Chapter 8—Comparison of Environment Map vs usage of a Background Plane with mapped texture*

Figure C–15b *Appendix A—Additonal Exercise for Chapter 5: Make better grass*

Figure C–16a *Bonus Excerpts—Renderings generated in the **Working with Modifiers (and more)** tutorial PDF: provided with the files from the Mastering VIZ Render CD ROM. Look for the MVIZr-Modifier.pdf PDF in the C:\MasterVIZr folder*

Figure C–16b *Bonus Excerpts—Renderings generated in the **Creating Signage** tutorial PDF: provided with the files from the Mastering VIZ Render CD ROM. Look for the MVIZr-Signage.pdf PDF in the C:\MasterVIZr folder*

4. Save and Close all files.

You can also add additional items and entourage to the model if you wish. Visit www.paulaubin.com for a tutorial explaining how the carriage lights were created. Feel free to add other items such as additional hardware, street furniture, mailboxes, or a fire hydrant. If you don't have models for these items, you can try your hand at building them yourself, or visit one of the many sites listed in Appendix B to search for both free and fee-based content to download.

CHAPTER 5

TOWNHOUSE

Here are some additional exercises to practice topics covered in Chapter 5 – Exterior Daylight Rendering.

Render Alternate Camera Angles

Render left – looking up.

1. From **Tools > Manage Scene States**, choose to **Save** your current scene state.

 NOTE From the VIZ Render User Reference:

Scene States offer a fast way to save different scene conditions with various lighting, camera, material, environment, and object properties that can be restored at any time and rendered to produce numerous interpretations of a model."

Try using this feature every time you change cameras and begin to set up a new rendering of the same VIZ Render file.

2. Right-click on a viewport label and choose **Views > Left – Looking Up** from the list.

3. Render the scene.

 NOTE You may need to modify the Height and/or Width for the output size before you render.

Render center – camera correction.

4. Right-click on a viewport label and choose **Views > Center – Camera Correction** from the list.

5. Render the scene.

 NOTE You may need to modify the Height and/or Width for the output size before your render.

Make Better Grass

The grass in front of the townhouse could use a little more realism.

6. Use the Selection Floater to select all the Mass Elements that make up the grass in the scene (*gra should get them)

7. In a zoomed in front or perspective view, move the grass up an inch or so.

8. Open Content Browser then the Render Material Catalog and then the Sitework > Planting category.

9. Drag Sitework.Planting.Grass.Thick and drop it onto your grass

10. Locate this new grass material in the Scene - In Use palette, right click and choose Properties.

11. Change the following values:

▶ Bump **500**

▶ Reflectance Scale **50**

▶ Indirect Bump Scale **40**

Click on the Bump Slot and change:

▶ Brightness **2**

12. Render scene with new material

Animate the Sun

Set up for the animation of the sun.

Do this exercise after you have completed Chapter 9.

Choose your favorite camera angle for this scene or create a new one.

13. Right-click on the Play button and adjust your frames to suit your desired output.

This will open a configuration window where you can set the desired length of your animation. Complete details are found in Chapter 9.

14. Select the Daylight system, and modify the date and time to the desired start of your animation.

15. Click the Auto button, move your slider to the last frame, and modify the sun's date and time to match the desired end of your animation.

16. Click the Auto button to finish.

17. Open the Render Scene dialog, click the lock icon next to the Image Aspect, and change either the Width or Height so that the Height is no more than **480** pixels.

 NOTE Most televisions do not display more than 480 lines vertically, unless you are rendering for HDTV, which is from 780 to 1,020 lines. Therefore rendering an animation to higher resolution than that is unnecessary and time-consuming.

Leave the Time Output at Single for running test renders.

 NOTE Keep in mind that Radiosity will run for each frame because the light is changing with each frame. With this in mind, lower Initial Quality with higher Filtering values will make rendering each frame a little faster. Also, Regather is extremely time-consuming, and is not recommended for animations.

18. When you are happy with the test renderings, save your scene.

You can closely approximate the total animation time required by the following formula:

Single Frame Render Time x Frames in Animation = Total Render Time.

19. In the Render dialog, set the Time Output to **Active Time Segment**.

20. Set to render individual frames via the Files button.

 TIP Create a folder to save the frames to.

21. Render the Animation.

22. When complete, compile the animation via RAM Player.

Complete details on the use of RAM Player can be found in Chapter 9.

CHAPTER 6

LIVING ROOM

Here are some additional exercises to practice topics covered in Chapter 6 – Interior Daylight Rendering.

Render Nighttime Scene

Set up for night shot.

1. Open the *Living Room.drf* and Save Copy As *Living Room Night.drf.* From the File menu, choose *Living Room Night* from the recent files list.

2. Select the Sun, and uncheck the On option under the Sun Parameters and IES Sky parameters rollouts in the Modify Panel. Verify that all the interior lights are On.

3. Open the Render Scene dialog to the Environment tab, and uncheck Use Map and change the Background Color to a dark blue.

Or, as an alternative, search the web for a nice nighttime photo similar to the ocean view used in the daytime rendering and load it instead.

4. In the Exposure Control rollout, click the Render Preview button. Adjust the Brightness if the Preview is washed out or too dark.

5. Render the Scene.

Render Interior Panorama

Set up and render a Panorama.

Do this exercise after you have completed Chapter 9.

6. Open either *Living Room* or *Living Room Night.*

7. Create a new camera in the center of the Living Room.

8. Adjust settings in the Render Scene dialog or continue to the next step if you are happy with your final render settings for this chapter.

9. On the Utility Panel, choose Panorama Exporter, and then choose Render.

10. Click the 1,024 x 512 Preset.

11. Render the Scene.

12. Open the Panorama in the Panoramic Viewer and click and drag your mouse.

Complete details on the use of the Panoramic Exporter can be found in Chapter 9.

CHAPTER 7

TOWNHOUSE

Here are some additional exercises to practice topics covered in Chapter 7 – Exterior Nighttime Rendering.

Render Alternate Camera Angles

Render left – looking up

1. From **Tools > Manage Scene States**, choose to **Save** your current scene state.

   NOTE Save a scene state for each different camera view in the following steps If you change renders settings for the camera vies, be sure to save a render preset for each; they will come in hand at the end of this section.

2. Right-click on a viewport label and choose **Views > Left – Looking Up** from the list.

   NOTE You may need to modify the Height and/or Width for the output size before you render.

Render center – camera correction

3. Right-click on a viewport label and choose **Views > Right – Cropped Render** from the list.

   NOTE You may need to modify the Height and/or Width for the output size before you render.

   From **Rendering > Batch Render**, **Add** three *renders* to be performed.

   NOTE Be sure to set what Camera, Scene state, and render Preset to use for each rendering. You may want to choose a "Draft"-type presets until you gain comfort with this feature.

   When you are finished, press **Render**.

CHAPTER 8

CONFERENCE ROOM

Here are some additional exercises to practice topics covered in Chapter 8 – Interior Space with Artificial Light.

Render Daytime Scene

Add RPC content.

1. Save the Scene as a new name and then open the new scene.

2. From the RPC menu, choose **RPC**.

If you have not installed the RCP Plugin, see the "Third Party Plug-ins" topic in Chapter 2 for information on how to install it.

3. Feel free to add any of the content from the list to your scene.

For example, you could place a laptop on the table. RPCs move and respond to the modify commands just like any other object in your scene. However, it is not recommended that you make clones (copies, references, or instances) of RPCs. If you need a second laptop, drop in another one from the modify panel instead.

4. Render the scene.

5. Save the Scene as a new name, modify the camera angle, and then render again.

Notice the way the RPC content responds to the change.

Set up for daytime shot.

6. Open the *Conference Room.drf* that you saved at the end of the chapter.

7. Save Copy As *Conference Room Day.drf*. From the File menu, choose *Conference Room Day* from the recent files list.

8. Add a daylight system and set the City to **Chicago**. Set the time to **11:00 AM** and the date to **February 29, 2005**.

This is the time and day that the photograph referenced in the next step was shot.

9. Right-click on the Skyline material in the Scene - In Use palette and choose **Copy**. On the Scene - Unused palette, right-click and choose **Paste**. Right-click on the copy and choose **Properties**.

10. Rename it **MVIZr Skyline - Day**. Click the Diffuse map button and then the Bitmap button. Double-click Bitmap and browse to your *C:\MasterVIZr\Maps* folder to select *Chicago-Day.jpg*. Close all dialogs to return to the scene.

11. Drag that new material to the background plane geometry.

12. Set the Skyline camera current.

13. Open the Render Scene dialog, and in the Exposure Control rollout, click the Render Preview button. Choose the Exterior Daylight checkbox.

    Adjust the Brightness if the Preview is washed out or too dark.

14. Place a checkmark in the Desaturate Low levels checkbox.

Since the background plane will be lit by the sun, this setting will help reduce the glowing effect of some of the colors in the skyline.

15. Render the Scene.

CHAPTER 9

Here are some additional exercises to practice topics covered in Chapter 9 – Animation and Client Presentation.

Render with Entourage

Add RPC content.

1. Open *Retreat – Animation Study.drf*.

2. Save it as *Retreat – Entourage.drf*, and then open it from the Recent files list.

3. From the RPC menu, choose **RPC**.

4. Add trees and shrubs to your scene. Keep in mind that Zero Z is at the floor of the building. Use the other viewports to locate your entourage on the surrounding landscape. This will be approximately –5'-0' in the Z.

 Note You can use color plates C–12 and C–13 as a guideline if you like.

5. Here again we will need to adjust the RPC's Self-Illumination due to the intensity of the sun. Select one RPC, and in the Modify Panel choose Mass Edit. In the Mass Edit dialog, select all the RPCs and adjust their Self-Illumination value to **100**, and click Apply.

6. When you are happy with your configuration, set the **Camera – Front Perspective** current.

7. Render the scene.

8. Repeat any of the animation exercises in Chapter 9 with the RPC entourage in place.

CONCLUSION AND SUMMARY

There are lots of other experiments that you can try. Please feel free to open any dataset from any chapter and explore and experiment. When you are ready, try some of the techniques covered in this book on your own projects. It is often good to make your first "real" project one that has already been completed by your firm. You can re-create the existing project quickly without the burden of making crucial design decisions at the same time that you are learning your new rendering skills. After this test project, you can move on to a real project that is already in or about to enter production.

Now that you have completed this book, dive in and have fun creating stunning presentations with ADT and VIZ Render!

APPENDIX b

Online Resources

In this appendix are listed several web sites and other resources that you can visit for information on VIZ Render, ADT, and related topics. All of these URLs are also available as shortcut files in the *Appendix B* folder on the CD ROM. You can simply browse these sites in your browser, or you can drop the entire contents into your "Favorites" folder for access via your web browser.

WEB SITES RELATED TO THE CONTENT OF THIS BOOK

www.paulaubin.com

Web site of the author. Includes information on this book and Aubin's other books, such as the *Mastering Autodesk Architectural Desktop* 2005 and 2004, as well as Aubin's two Release 3.3 books: *Mastering Autodesk Architectural Desktop R3* and *Autodesk Architectural Desktop: An Advanced Implementation Guide.* Check here for ordering information and addenda. You can also find information on training and consulting services offered by the author.

Email Paul F. Aubin or James D. Smell at MasterVIZr@paulaubin.com.

www.autodeskpress.com

The web site for Autodesk Press. Visit for information on other CAD titles, on-line resources, student software, and more.

www.autodesk.com

Autodesk's main Web site. Visit often for the latest information on Autodesk products.

www.archvision.com

With more than 70 collections of automobiles, people, trees, and other objects, ArchVision is the leading image-based content supplier in the design visualization industry. Visit www.archvision.com for demos and samples.

www.hermanmiller.com

Browse HermanMiller.com to learn more about the company, products, the research and the designers behind them. Also find complete information on the services provided by Herman Miller as well as locate your local dealer and retailer contact information.

www.scottonstott.com

Scott is an architectural software guru who publishes video courses on AutoCAD, Revit, 3ds max, VIZ, and Photoshop. He has contributed to over a dozen books and has created numerous visualizations for AEC firms. The site offers free downloads of quality designer furniture 3D models, AutoLISP and MAXscripts, a video demo, and much more.

www.vizdepot.com

The Vizdepot is your resource for all of the latest information for professional visualization artists and architects. The site includes support of all releases of VIZ, including ADT 2004/VIZ Render, ADT 2005/VIZ Render 2005, and 3ds max 6. Registration with this web site enables you to post to the forums, add images to the galleries, as well as access the many new tutorials and contents of the site. The site also includes textures, productivity tips, downloads, scripts, and more.

WEB SITES OF RELATED INTEREST

discussion.autodesk.com

Autodesk Discussion Groups main page. Online community of Autodesk users sharing comments, questions, and solutions about all Autodesk products.

autodesk.blogs.com/between_the_walls/

Chris Yanchar, a product designer on the Architectural Desktop team, shares and aggregates information on Architectural Desktop and architecture. At this Blog site you will also find links to How To's and articles about ADT features.

autodesk.blogs.com/between_the_lines/

Shaan Hurley, a Technical Marketing Manager for the AutoCAD group and manager of most of the beta programs for Autodesk, shares his views on AutoCAD, technology, and life. Not everything is official Autodesk opinion, endorsement, or recommendation—this is a Blog from Shaan himself. This approach provides direct contact between customers and Autodesk personnel most familiar with the products.

www.discreet.com/

www.siggraph.org/education/curriculum/projects/slide_sets/slides93/01_93_3.htm

www.siggraph.org/education/curriculum/projects/slide_sets/slides91/91_01_0.htm

www.siggraph.org/education/materials/HyperGraph/raytrace/rtrace0.htm

CADalyst Magazine web forum

cadence.advanstar.com//ubbthreads/ubbthreads.php?Cat=

Online user forum hosted by *CADalyst Magazine* and moderated by author Paul F. Aubin.

www.cadalyst.com/cadalyst/

Main home page for *CADalyst Magazine.* View magazines online or subscribe to the print edition.

www.aia.org

Web site of the American Institute of Architects.

MATERIAL IDEAS:

www.3-form.com

www.antolini.it

www.beldenbrick.com

cgarchitect.com

www.daviltravertini.com

www.benjaminmoore.com

www.animax.it/

www.carolinaceramics.com/

www.mimosainternational.com/

www.ambientlight.co.uk/

astronomy.swin.edu.au/~pbourke/texture/

pchan.cgworks.com/tutorials/

www.viz2000.com/html/alfa/materials.html

LIGHTING:

www.louispoulsen.com

www.eLumit.com

www.erco.com

www.holophane.com/

www.ledalite.com/

DIGITAL OUTPUT:

www.tdp.nu/rbenchmarks/

www.audiovideo101.com/dictionary/hdtv.asp

developer.msntv.com/Designing/tvsrnres.asp

members.aol.com/ajaynejr/vidres.htm

ENTOURAGE:

www.got3d.com/

www.itoosoft.com/

www.nsight3d.com/

MODELS:

www.evermotion.org/

www.3dcafe.com

www.turbosquid.com

www.3drender.com/light/index.html

www.e-interiors.net

www.mr-cad.com/default.php

www.ultra3d.com/

RENDERING COMMUNITY:

www.blur.com/indexl.html

www.kdlab.net/

www.arquiserveis.com/

www.origamy.com.br/

www.tbmax.com/

www.ubikmh.nl/

rndr4food.blogspot.com/

APPENDIX C

Render Settings

INTRODUCTION

In this Appendix, we present three tables. Each of these tables presents a tab of the Render Scene dialog and shows how the various settings interact and/or contribute to the quality of the rendering. This Appendix is also provided in digital format on the CD. You can access it by choosing the **Install Appendix C Files** button on the CD ROM install screen. See the Preface for more information.

RENDER SCENE DIALOG

COMMON TAB

Common Parameters

Time Output						
Single	N	-	-	-	-	-
Active Time Segment:	N	-	-	-	-	-
Range:	N	-	-	D	-	-
Frames	N	-	-	-	-	-
File Number Base:	N	-	-	-	-	-
Every Nth Frame:	N	-	-	-	-	-
Output Size						
Width	N	-	-	D x D	I	32768
Height	N	-	-	D x D	I	32768
Image Aspect	N	-	-	-	0.001	1000

Rollout Group Setting	Invalidates Radiosity	Higher or Lower value increases quality	ON or OFF speeds Render time	Directly or Exponentially affects Render time	Min Value	Max Value
Render Output						
Save File	N	-	-	-	-	-
Rendered Frame Window	N	-	-	-	-	-
Skip Existing Images	N	-	-	-	-	-
Photorealistic Rendering						
Anitalias Geometry	N	-	OFF	-	-	-
Mapping	Y	-	OFF	-	-	-
Shadow/Material Quality	N	H	-	E	0	100
Shadows	Y	-	OFF	-	-	-
Render Options						
Reflections/Refractions	Y	-	OFF	E	-	-
Effects	N	-	OFF	-	-	-
Use Radiosity	Y	-	ON	-	-	-
Advanced Options						
Two-Sided Materials	Y	-	OFF	D	-	-
Fast Linear/Area Shadows	Y	-	ON	-	-	-
Compute Radiosity when Required	Y	-	OFF	-	-	-
Two-Sided Shadows	Y	-	OFF	-	-	-
Ray Bounces	N	H	-	E	0	50
Watermark Parameters						
Render	N	-	-	-	-	-
Blend value	N	-	-	-	0.0	1.0
Image Top	N	-	-	-	0	32768
Image Left	N	-	-	-	0	32768
Assign Renderer						
Current Renderers						
Renderer	Y	-	-	-	-	-

RADIOSITY
TAB
Radiosity Processing Parameters

Process

Initial Quality	N	H	-	D	0	100
Refine Iterations (All Objects)	N	H	-	D	0	10000
Refine Iterations (Selected Objects)	N	H	-	-	0	10000
Process Refine Iterations Stored in Objects	N	-	-	-	-	-
Update Data When Required on Start	Maybe	-	-	-	-	-

Interactive Tools

Indirect Light Filtering	N	L	-	-	0	1000
Direct Light Filtering	N	L	-	-	0	1000
Logarithmic Exposure Control	N	-	-	-	-	-
Display Radiosity in Viewport	N	-	-	-	-	-

Radiosity Meshing Parameters

Global Subdivision Settings

Enabled	Y	-	-	-	-	-
Use Adaptive Subdivision	Y	-	-	-	-	-

Mesh Settings

Maximum Mesh Size	Y	L	-	E	.01m/ 3/8"	10000m/ 32808ft
Minimum Mesh Size	Y	L	-	E	.01m/ 3/8"	10000m/ 32808ft
Contrast Threshold	Y	L	-	E	0	100
Initial Mesh Size	Y	L	-	E	.01m/ 3/8"	10000m/ 32808ft

Light Settings

Shoot Direct Lights	Y	-	-	-	-	-
Include Point Lights in Subdivision	Y	-	-	-	-	-
Include Linear Lights in Subdivision	Y	-	-	-	-	-
Include Area Lights in Subdivision	Y	-	-	-	-	-
Include Skylight	Y	-	-	-	-	-
Include Self-Emitting Faces in Subdivision	Y	-	-	-	-	-

Rollout Group Setting	Invalidates Radiosity	Higher or Lower value increases quality	ON or OFF speeds Render time	Directly or Exponentially affects Render time	Min Value	Max Value
Minimum Self-Emitting Size	Y	L	-	-	.01m/ 3/8"	10000m/ 32808ft
Light Painting						
Intensity	N	-	-	-	0	10000
Pressure	N	-	-	-	0	100
Rendering Parameters						
Re-Use Direct Illumination from Radiosity Solution	N	-	-	-	-	-
Render Direct Illumination	N	-	-	-	-	-
Regather Indirect Illumination						
Regather	N	-	OFF	-	-	-
Rays per Sample	N	H	-	E	1	10000
Filter Radius (pixels)	N	L	-	D	0.5	30
Clamp Values	N	-	-	-	0	1000000
Adaptive Sampling						
Initial Sample Spacing	N	L	-	-	2x2	32x32
Subdivision Contrast	N	L	-	D	0	200
Subdivide Down To	N	L	-	-	1x1	32x32
Show Samples	N	-	-	-	-	-
Statistics						

ENVIRONMENT TAB
Background
Background

	Invalidates Radiosity	Higher or Lower value increases quality	ON or OFF speeds Render time	Directly or Exponentially affects Render time	Min Value	Max Value
Color	N	-	-	-	-	-
Environment Map	N	-	-	-	-	-

Rollout Group Setting	Invalidates Radiosity	Higher or Lower value increases quality	ON or OFF speeds Render time	Directly or Exponentially affects Render time	Min Value	Max Value
Use Map	N	-	-	-	-	-
Ambient	N	-	-	-	-	-
Background Control						
Scale	-	-	-	-	0.1	100
Offset U:	-	-	-	-	-10	10
Offset V:	-	-	-	-	-10	10
Ambient	-	-	-	-	-	-
Exposure Control						
Type	N	-	-	-	-	-
Active	N	-	-	-	-	-
Exposure Control Parameters						
Automatic Exposure Control						
Brightness	N	-	-	-	0	100
Contrast	N	-	-	-	0	100
Exposure Value	N	-	-	-	0	5
Physical Scale	N	-	-	-	0.001	200000
Color Correction	N	-	-	-	-	-
Desaturate Low Levels	N	-	-	-	-	-
Logarithmic Exposure Control						
Brightness	N	-	-	-	0	200
Contrast	N	-	-	-	0	100
Mid Tones	N	-	-	-	0.01	20
Physical Scale	N	-	-	-	0.001	200000
Color Correction	N	-	-	-	-	-
Desaturate Low Levels	N	-	-	-	-	-
Affect Indirect Only	N	-	-	-	-	-
Exterior Daylight	N	-	-	-	-	-

Rollout Group Setting	Invalidates Radiosity	Higher or Lower value increases quality	ON or OFF speeds Render time	Directly or Exponentially affects Render time	Min Value	Max Value
Pseudo Color Exposure Control						
Quantity	N	-	-	-	-	-
Style	N	-	-	-	-	-
Scale	N	-	-	-	-	-
Min	N	-	-	-	0	200000
Max	N	-	-	-	0	200000
Physical Scale	N	-	-	-	0.001	200000

RenderTimes

INTRODUCTION

In this Appendix, we have listed the time it took for each rendering on both Paul's and Jim's machines. Included in the table are the Rendering and Radiosity Processing times (as appropriate) for each system. In addition, we have reiterated the music selection choices and color plates for renderings included in the color section. All of this information is also included inline in the text of each chapter.

PAUL'S MACHINE

Paul is using a Dell Inspiron 8500 laptop. It has a Mobile Intel® Pentium® 4-M CPU running at 2.20GHz. His system has: 1.25 GB of RAM. He is running Microsoft Windows XP Professional, Version 2002, Service Pack 1. Paul has a 60 GB Hard Drive and a NVIDIA GeForce4 4200 Go (Dell Mobile) video card.

JIM'S MACHINE

Jim is using a custom-built desktop. It has Dual Intel® Pentium® 3 CPUs running at 850MHz. His system has: 512 MB of RAM. He is running Microsoft Windows 2000, Service Pack 4. Jim has a 120 GB and a 100 GB Hard Drive and an ATI Radeon 7200 video card.

 Note: Both Jim's and Paul's machines would be considered lower mid-range systems. If you have a system that was purchased in the last 12 to 18 months, there is a very good chance that your render times will be considerably better than those listed here.

RENDER STATISTICS

You can find a list or Render Times with thumbnail images at: *http://www.paulaubin.com/mastering_viz.php*. Render stats are provided as a guide. Your times will vary with your specific hardware specifications.

	Paul's		Jim's			
	Render Time	Radiosity Time	Render Time	Radiosity Time	Music Selection	Color Plate
Chapter 1						
Render 01.01	0:01:26	NA	0:00:58	NA	"Clouds Up" Air – The Virgin Suicides [Original Soundtrack] (2000)	Rendering 01.01.jpg
Render 01.02	0:00:31	NA	0:00:06	NA	Final Jeopardy Clock Music	Rendering 01.02.jpg
Render 01.03	0:00:21	0:00:06	0:00:20	0:00:13	"Stop" Pink Floyd – The Wall Disk 2 (1979)	Rendering 01.03.jpg
Render 01.04	0:00:13	0:00:04	0:00:22	0:00:15	"Rally" Prince of Egypt – Soundtrack (1998)	Rendering 01.04.jpg
Render 01.05	0:05:54	0:04:14	0:07:30	0:01:41	"Blue Monday" New Order – Substance 1987 (Disc 1)	Rendering 01.05.jpg
Render 01.06			0:25:11	0:01:44	The Cars – The Cars (1978) Entire CD	
Render 01.07			03:30:47	0:05:54	Watch "The Godfather (1972) Part I" on DVD	C–1
Chapter 2						
Render 02.01	0:00:11	NA	0:00:13	NA	"Love Me Do" The Beatles – The Beatles 1 (2000)	Rendering 02.01.jpg
Render 02.02	0:00:13	0:00:04	0:00:15	0:00:03	"Butterfly" Lenny Kravitz – Mama Said (1991)	Rendering 02.02.jpg
Render 02.03	0:01:18	0:00:53	0:02:12	0:00:47	"Don't Know Why" Norah Jones – Come Away With Me (2002)	C–3a
Render 02.04			2:48:22	0:00:18	Watch Episode 1, the Pilot of "Smallville - The Complete First Season (2001)" on DVD	C–3b
Chapter 5						
Render 05.01	0:00:34		0:00:45		"Symphony No. 9 (Scherzo)" Ludwig van Beethoven, composer. Seattle Symphony. Gerard Schwarz, director – Beethoven's Symphony No. 9 (Scherzo)	Rendering 05.01.jpg
Render 05.02	0:00:36		0:00:52		"Chicago" Frank Sinatra, Nelson Orchestra Riddle – The Capitol Years: The Best of Frank Sinatra Disc 2 (1953)	Rendering 05.02.jpg
Render 05.03	0:00:42		0:01:02		"Young Americans [Single Version]" David Bowie – Best of Bowie [Virgin US/Canada] (2002)	Rendering 05.03.jpg
Render 05.04	0:00:49		0:00:46		"In the End" Linkin Park – Hybrid Theory (2000)	Rendering 05.04.jpg
Render 05.05	0:00:51		0:00:47		"Beth" Kiss – The Very Best of Kiss (2002)	Rendering 05.05.jpg

	Paul's		Jim's			
	Render Time	**Radiosity Time**	**Render Time**	**Radiosity Time**	**Music Selection**	**Color Plate**
Render 05.06	0:00:27		0:00:30		"Fotografía" Juanes; Nelly Furtado – Un Dia Normal (2002)	Rendering 05.06.jpg
Render 05.07	0:00:42		0:00:45		"MacArthur Park" Donna Summer – On the Radio - Greatest Hits Vols. I and II (1987)	Rendering 05.07.jpg
Render 05.08	0:08:52	0:07:52	0:12:58	0:07:18	"The Dark Knight" Branford Marsalis Quartet – Crazy People Music	C–5
Chapter 6						
Render 06.01	0:02:11	0:01:22	0:02:20	0:01:30	"Panic" The Smiths – Singles	Rendering 06.01.jpg
Render 06.02	0:02:26	0:01:37	0:02:15	0:01:30	"Tush" ZZ Top – Fandango	Rendering 06.02.jpg
Render 06.03	0:01:26	0:01:17	0:02:06	0:01:21	"In Trutina" Charlotte Church – Voice of an Angel	Rendering 06.03.jpg
Render 06.04	0:01:30	0:01:20	0:01:33	0:01:21	"Tanz Der Schwane" Tchaikovsky – Swan Lake	Rendering 06.04.jpg
Render 06.05	0:01:22	0:01:13	0:01:34	0:01:21	"Coming Clean" Green Day – Dookie	Rendering 06.06.jpg
Render 06.06	0:01:35	0:01:24	0:01:35	0:01:22	"The Ocean" U2 – Boy	Rendering 06.06.jpg
Render 06.07	0:01:37	0:01:26	0:01:39	0:01:26	"Winter" John Denver – Portrait (Disc 1)	Rendering 06.07.jpg
Render 06.08	0:13:07	0:09:37	0:06:42	0:04:08	"Take My Hand" Dido – No Angel	Rendering 06.08.jpg
Render 06.09	0:02:42		0:02:54		"She's A Lady" Tom Jones – The Best of…Tom Jones	Rendering 06.09.jpg
Render 06.10			11:56:32	0:12:17	Watch: "Gone with the Wind (1939)" and then "Casablanca (Two-Disc Special Edition) (1942)" on DVD	C–6
Chapter 7						
Render 07.01	0:13:14	0:12:51	0:11:06	0:10:40	"All Blues" Miles Davis – Kind of Blue (Remastered) (1997)	Rendering 07.01.jpg
Render 07.02	0:15:14	0:14:36	0:12:21	0:11:21	"Piano Concerto In D Minor, K.466: 1st Movement" Neville Marriner – Amadeus Original Soundtrack Vol 2.	Rendering 07.02.jpg
Render 07.03	0:46:22	0:19:53	0:34:28	0:15:09	Gipsy Kings – Gipsy Kings (1988) Entire CD	Rendering 07.03.jpg
Render 07.04	0:46:22	0:19:53	0:34:28	0:15:09	Fleetwood Mac – Rumours (1977) Entire CD	Rendering 07.04.jpg
Render 07.05			8:25:48	1:12:32	Watch the anxiously awaited "Star Wars Trilogy (Widescreen Edition)" on DVD (2004)	C–8

	Paul's		Jim's			
	Render Time	Radiosity Time	Render Time	Radiosity Time	Music Selection	Color Plate
Chapter 8						
Render 08.01	0:04:47	0:03:08	0:05:22	0:03:53	"Family Snapshot" Peter Gabriel – Shaking the Tree: Sixteen Golden Greats (1990)	Rendering 08.01.jpg
Render 08.02	0:19:12	0:15:41	0:22:00	0:18:58	"Clocks" Coldplay – A Rush of Blood to the Head (2002)	Rendering 08.02.jpg
Render 08.03			8:03:23	0:19:29	Watch "Office Space (Widescreen Edition)" on DVD (1999)	C–10
Render 08.04			9:13:50	0:25:21	Watch The Lord of the Rings - The Fellowship of the Ring (Platinum Series Special Extended Edition) (2001) on DVD	C–11
Chapter 9						
Render 09.01	0:00:39	0:00:14	0:00:45	0:00:20	"Driven Under" Seether – Disclaimer (2002)	Rendering 09.01.jpg
Render 09.02	0:06:31 total (0:00:04 per Frame)	0:00:16	0:12:00	0:00:20	"Bohemian Rhapsody" Queen – Greatest Hits, Vols. 1-2 Disc 1 (1995)	Animation 09.02.avi and C12
Render 09.03	0:06:15 total (0:00:04 per Frame)	0:00:06	0:08:20	0:00:10	"Bad" U2 – The Unforgettable Fire (1984)	Animation 09.03.avi and C–14 bottom
Render 09.04	0:00:45	0:00:20	0:03:00	0:00:20	"Forever And Always" Shania Twain – Up! (2002)	Rendering 09.04.jpg
Render 09.05			0:06:00 total (0:01:00 per Frame)	0:4:07	"Rapture" Blondie – The Best Of Blondie (1989)	Rendering 09.05.jpg
Render 09.06			0:30:00 total (0:5:00 per Frame)	0:4:07	"Cling" Days Of The New – Days Of The New (1997)	Rendering 09.06.jpg and C–14 middle
Render 09.07			0:25:15 total (0:0:16 per Frame)	0:20:00 total (0:0:12 per Frame)	Eric Clapton - The Cream of Clapton, (1995) Entire CD	Animation 09.07 .avi and C–13

INDEX

LICENSE AGREEMENT FOR AUTODESK PRESS

A Thomson Leaning Company

IMPORTANT-READ CAREFULLY: This End User License Agreement ("Agreement") sets forth the conditions by which Delmar Learning, a division of Thomson Learning Inc. ("Thomson") will make electronic access to the Thomson Delmar Learning-owned licensed content and associated media, software, documentation, printed materials and electronic documentation contained in this package and/or made available to you via this product (the "Licensed Content"), available to you (the "End User"). BY CLICKING THE "I ACCEPT" BUTTON AND/OR OPENING THIS PACKAGE, YOU ACKNOWLEDGE THAT YOU HAVE READ ALL OF THE TERMS AND CONDITIONS, AND THAT YOU AGREE TO BE BOUND BY ITS TERMS CONDITIONS AND ALL APPLICABLE LAWS AND REGULATIONS GOVERNING THE USE OF THE LICENSED CONTENT.

1.0 SCOPE OF LICENSE

1.1 Licensed Content. The Licensed Content may contain portions of modifiable content ("Modifiable Content") and content which may not be modified or otherwise altered by the End User ("Non-Modifiable Content"). For purposes of this Agreement, Modifiable Content and Non-Modifiable Content may be collectively referred to herein as the "Licensed Content." All Licensed Content shall be considered Non-Modifiable Content, unless such Licensed Content is presented to the End User in a modifiable format and it is clearly indicated that modification of the Licensed Content is permitted.

1.2 Subject to the End User's compliance with the terms and conditions of this Agreement, Thomson Delmar Learning hereby grants the End User, a nontransferable, non-exclusive, limited right to access and view a single copy of the Licensed Content on a single personal computer system for noncommercial, internal, personal use only. The End User shall not (i) reproduce, copy, modify (except in the case of Modifiable Content), distribute, display, transfer, sublicense, prepare derivative work(s) based on, sell, exchange, barter or transfer, rent, lease, loan, resell, or in any other manner exploit the Licensed Content; (ii) remove, obscure or alter any notice of Thomson Delmar Learning's intellectual property rights present on or in the License Content, including, but not limited to, copyright, trademark and/or patent notices; or (iii) disassemble, decompile, translate, reverse engineer or otherwise reduce the Licensed Content.

2.0 TERMINATION

2.1 Thomson Delmar Learning may at any time (without prejudice to its other rights or remedies) immediately terminate this Agreement and/or suspend access to some or all of the Licensed Content, in the event that the End User does not comply with any of the terms and conditions of this Agreement. In the event of such termination by Thomson Delmar Learning, the End User shall immediately return any and all copies of the Licensed Content to Thomson Delmar Learning.

3.0 PROPRIETARY RIGHTS

3.1 The End User acknowledges that Thomson Delmar Learning owns all right, title and interest, including, but not limited to all copyright rights therein, in and to the Licensed Content, and that the End User shall not take any action inconsistent with such ownership. The Licensed Content is protected by U.S., Canadian and other applicable copyright laws and by international treaties, including the Berne Convention and the Universal Copyright Convention. Nothing contained in this Agreement shall be construed as granting the End User any ownership rights in or to the Licensed Content.

3.2 Thomson Delmar Learning reserves the right at any time to withdraw from the Licensed Content any item or part of an item for which it no longer retains the right to publish, or which it has reasonable grounds to believe infringes copyright or is defamatory, unlawful or otherwise objectionable.

4.0 PROTECTION AND SECURITY

4.1 The End User shall use its best efforts and take all reasonable steps to safeguard its copy of the Licensed Content to ensure that no unauthorized reproduction, publication, disclosure, modification or distribution of the Licensed Content, in whole or in part, is made. To the extent that the End User becomes aware of any such unauthorized use of the Licensed Content, the End User shall immediately notify Delmar Learning. Notification of such violations may be made by sending an Email to delmarhelp@thomson.com.

5.0 MISUSE OF THE LICENSED PRODUCT

5.1 In the event that the End User uses the Licensed Content in violation of this Agreement, Thomson Delmar Learning shall have the option of electing liquidated damages, which shall include all profits generated by the End User's use of the Licensed Content plus interest computed at the maximum rate permitted by law and all legal fees and other expenses incurred by Thomson Delmar Learning in enforcing its rights, plus penalties.

6.0 FEDERAL GOVERNMENT CLIENTS

6.1 Except as expressly authorized by Delmar Learning, Federal Government clients obtain only the rights specified in this Agreement and no other rights. The Government acknowledges that (i) all software and related documentation incorporated in the Licensed Content is existing commercial computer software within the meaning of FAR 27.405(b)(2); and (2) all other data delivered in whatever form, is limited rights data within the meaning of FAR 27.401. The restrictions in this section are acceptable as consistent with the Government's need for software and other data under this Agreement.

7.0 DISCLAIMER OF WARRANTIES AND LIABILITIES

7.1 Although Thomson Delmar Learning believes the Licensed Content to be reliable, Thomson Delmar Learning does not guarantee or warrant (i) any information or materials contained in or produced by the Licensed Content, (ii) the accuracy, completeness or reliability of the Licensed Content, or (iii) that the Licensed Content is free from errors or other material defects. THE LICENSED PRODUCT IS PROVIDED "AS IS," WITHOUT ANY WARRANTY OF ANY KIND AND THOMSON DELMAR LEARNING DISCLAIMS ANY AND ALL WARRANTIES, EXPRESSED OR IMPLIED, INCLUDING, WITHOUT LIMITATION, WARRANTIES OF MERCHANTABILITY OR FITNESS OR A PARTICULAR PURPOSE. IN NO EVENT SHALL THOMSON DELMAR LEARNING BE LIABLE FOR: INDIRECT, SPECIAL, PUNITIVE OR CONSEQUENTIAL DAMAGES INCLUDING FOR LOST PROFITS, LOST DATA, OR OTHERWISE. IN NO EVENT SHALL DELMAR LEARNING'S AGGREGATE LIABILITY HEREUNDER, WHETHER ARISING IN CONTRACT, TORT, STRICT LIABILITY OR OTHERWISE, EXCEED THE AMOUNT OF FEES PAID BY THE END USER HEREUNDER FOR THE LICENSE OF THE LICENSED CONTENT.

8.0 GENERAL

8.1 Entire Agreement. This Agreement shall constitute the entire Agreement between the Parties and supercedes all prior Agreements and understandings oral or written relating to the subject matter hereof.

8.2 Enhancements/Modifications of Licensed Content. From time to time, and in Delmar Learning's sole discretion, Thomson Thomson Delmar Learning may advise the End User of updates, upgrades, enhancements and/or improvements to the Licensed Content, and may permit the End User to access and use, subject to the terms and conditions of this Agreement, such modifications, upon payment of prices as may be established by Delmar Learning.

8.3 No Export. The End User shall use the Licensed Content solely in the United States and shall not transfer or export, directly or indirectly, the Licensed Content outside the United States.

8.4 Severability.If any provision of this Agreement is invalid, illegal, or unenforceable under any applicable statute or rule of law, the provision shall be deemed omitted to the extent that it is invalid, illegal, or unenforceable. In such a case, the remainder of the Agreement shall be construed in a manner as to give greatest effect to the original intention of the parties hereto.

8.5 Waiver. The waiver of any right or failure of either party to exercise in any respect any right provided in this Agreement in any instance shall not be deemed to be a waiver of such right in the future or a waiver of any other right under this Agreement.

8.6 Choice of Law/Venue. This Agreement shall be interpreted, construed, and governed by and in accordance with the laws of the State of New York, applicable to contracts executed and to be wholly preformed therein, without regard to its principles governing conflicts of law. Each party agrees that any proceeding arising out of or relating to this Agreement or the breach or threatened breach of this Agreement may be commenced and prosecuted in a court in the State and County of New York. Each party consents and submits to the non-exclusive personal jurisdiction of any court in the State and County of New York in respect of any such proceeding.

8.7 Acknowledgment. By opening this package and/or by accessing the Licensed Content on this Website, THE END USER ACKNOWLEDGES THAT IT HAS READ THIS AGREEMENT, UNDERSTANDS IT, AND AGREES TO BE BOUND BY ITS TERMS AND CONDITIONS. IF YOU DO NOT ACCEPT THESE TERMS AND CONDITIONS, YOU MUST NOT ACCESS THE LICENSED CONTENT AND RETURN THE LICENSED PRODUCT TO THOMSON DELMAR LEARNING (WITHIN 30 CALENDAR DAYS OF THE END USER'S PURCHASE) WITH PROOF OF PAYMENT ACCEPTABLE TO DELMAR LEARNING, FOR A CREDIT OR A REFUND. Should the End User have any questions/comments regarding this Agreement, please contact Thomson Delmar Learning at delmar-help@thomson.com.